Field Hydrogeology

The Geological Society of London Handbook Series
published in association with
the Open University Press comprises

Barnes: *Basic Geological Mapping*
Tucker: *The Field Description of Sedimentary Rocks*
Fry: *The Field Description of Metamorphic Rocks*
Thorpe and Brown: *The Field Description of Igneous Rocks*
McClay: *The Mapping of Geological Structures*
Milsom: *Field Guide to Geophysics* (in preparation)
Goldring: *Fossils in the Field* (in preparation)

Professional Handbooks in the Series

Brassington: *Field Hydrogeology*
Clark: *The Field Guide to Water Wells and Boreholes*

Geological Society of
London Professional Handbook

HANDBOOK SERIES EDITOR – M. H. de FREITAS

Field Hydrogeology

R. Brassington

Planning Department
North West Water
Warrington

OPEN UNIVERSITY PRESS
MILTON KEYNES
and
HALSTED PRESS
John Wiley & Sons
New York – Toronto

Open University Press
Open University Educational Enterprises Limited
12 Cofferidge Close
Stony Stratford
Milton Keynes MK11 1BY

First published 1988

GB
1002.3
B73
1988

British Library Cataloguing in Publication Data

Brassington, Rick
    Field hydrogeology. — (Geological Society
    of London handbook).
    1. Hydrogeology
    I. Title   II. Series
    551.49   GB1003.2

    ISBN 0-335-15202-3

Published in the U.S.A., Canada and Latin America by
Halsted Press, a Division of John Wiley & Sons, Inc., New York

Library of Congress Cataloging in Publication Data

Brassington, R. (Richard)
    Field hydrology.

    Bibliography: p.
    1. Hydrology — Field work.   I. Title.
GB658.4.B7   1988   551.48   87-38330

    ISBN 0-470-21078-8

Printed in Great Britain

To Sandra, David and Elizabeth

Contents

Preface

This book is primarily intended to help graduate geologists and others at an early stage in their professional career, or with limited practical experience in hydrogeology, to undertake the basic fieldwork necessary to understand the hydrogeology of an area. It is a field guide with an emphasis on the 'how to' aspects, rather than a textbook which addresses the question 'why?'. Hydrogeology is not entirely a field-based branch of the geological sciences, and consequently it has been necessary to include a certain amount of background theory and analytical methods to explain why particular field measurements are required, and the importance of making them in a particular way.

The techniques explained in this Handbook are essentially *basic*, but the reader should not assume that field hydrogeology excludes sophisticated methods. This is far from the case. Increasingly, electronic equipment collecting data in digital form, and computer-based methods for handling these data, are employed, but the value of the answers so obtained relies entirely upon the relevance of the information used, and this in turn relies upon a consideration of the fundamental geological controls on groundwater movement. This Handbook is intended to be a ready reference to simple field techniques which will help secure a knowledge of the fundamental controls operating in an area, and which can be used as a basis from which more elaborate theories may be developed.

Hydrogeologists are often employed as part of a team, but equally they are likely to work alone, perhaps in parts of the world where either access to back-up equipment or advice are not available. This Handbook is also intended to meet these needs by providing information on practical field techniques, including some useful 'tricks of the trade' which help to improvise work in the field, so that important measurements are not missed.

It is assumed that readers have a good understanding of geological principles and have undertaken geological fieldwork, probably undergraduate mapping. A variety of basic skills will be needed, including map reading, methodical recording of field observations, plotting graphs, drawing diagrams and report writing. An ability to drive, carry out running repairs on equipment, an aptitude for working alone (sometimes in remote areas) and a knowledge of first aid and survival skills are all likely to be important.

The order in which topics are covered in the book is designed to take the reader logically through the steps involved in a groundwater investigation in an unknown area. Early chapters cover the choice of equipment, collation of the available information and planning a fieldwork programme. Various field methods are then described, including the measurement of rainfall and surface water flow. Advice is offered on carrying out a wide range of specific investigations to help place the earlier chapters in an everyday context. The final chapter provides suggestions on safe working practices.

Acknowledgements

I would like to acknowledge the help and guidance given to me by a large number of people during the writing of this book, particularly Mike de Freitas – the Handbook Series Editor, and Rose Dixon of the Open University Press. I am grateful to my colleague Keith Seymour for his detailed criticism of the manuscript and to many other friends and colleagues who, consciously or unconsciously, have inspired parts of the text. Thanks are due to Dave Passey and Jim Campbell for help with the photographs, and to Ron Shaw for designing and building a working simple dipper. I am very grateful to my wife for her patience during the writing of this book, for her help in improving the text and for typing the final manuscript.

The following people provided information on data sources:
The Director, US Geological Survey, Keith Bliss, Dr K. Bloomfield, Donal Daly, Dr Wojtek Kawecki, Colin Patrick, Dr K. V. Raghava Rao, George Reeves, John Riden, Professor Ken Rushton and Rob Sage.

Illustrative matter in this book was adopted from or inspired by the following sources, to whom grateful acknowledgement is made:

Fig. 2.10: J. W. Barnes, *Basic Geological Mapping*, Open University Press (1981); Figs. 4.2, 6.9, 6.13, 6.18, 7.18, 7.20, 7.21: North-West Water field data; Figs. 5.2, 5.3: D. N. Cargo and B. F. Mallory, *Man and His Geologic Environment*, Addison-Wesley (1974); Table 5.1: D. A. Morris and A. I. Johnson, *US Geol. Surv. Water Supply Pap.* **1839-D** (1967); Table 5.2: A. I. Johnson, *US Geol. Surv. Water Supply Pap.* **1662-D** (1967); Table 5.3: Dept. of Economic and Social Affairs, *Groundwater in the Western Hemisphere*, United Nations (1976); Tables 5.4, 9.1, Fig. 6.16: The Open University, *S238 The Earth's Physical Resources – Block 4 Water Resources*, Open University Press (1984); Fig. 5.6: Bureau of Reclamation, *Groundwater Manual*, US Dept. Interior (1977); Fig. 6.10: Inst. of Hydrology, *Hydrological Data UK*, Natural Environment Research Council (1984); Table 7.4, Figs. 7.22, 7.23: D. P. Drew and D. I. Smith, *Techniques for the Tracing of Subterranean Drainage*, British Geomorphological Research Group (1969) Geobooks Ltd; Table 8.2: H. L. Penman, *Proc. R. Soc. Lond.*, **193** (1948).

1
Introduction

1.1 Why investigate groundwater?

Groundwater plays a very important role in many geological processes. The presence or absence of groundwater, its chemistry and temperature, may all significantly affect lithification, and groundwater is essential for hydrothermal processes and the genesis of many ore veins and bodies. In applied geology too, groundwater is important. The engineering properties of rocks and soils are often controlled by groundwater, and changes in groundwater conditions may have a disasterous effect on the stability of slopes, buildings or other structures. A study of groundwater conditions is, therefore, an essential element in a site investigation for new construction works. Perhaps the most important feature of groundwater, however, is its value as a resource for water supplies.

This Handbook is concerned with the field techniques used by hydrogeologists to evaluate groundwater systems. Such studies are needed to position new wells in favourable sites – a job still left to the water diviner in some countries – or to calculate the total quantity of groundwater available in an area. Pumping from new wells may reduce the quantities which can be pumped from others nearby, or cause local spring flows to dwindle; the hydrogeologist will be expected to make predictions on such effects and can only do so if he has a proper understanding of the local groundwater system based on adequate field observations. It is equally important to evaluate the quality of groundwater to ensure that it is suitable for drinking or for other uses. Hydrogeological studies are also needed to assess the hazards of waste disposal sites (landfill), septic tanks and other activities which may pollute groundwater resources and supplies. In such instances, detailed hydrogeological investigations may need to be carried out in a limited area, such as at the waste disposal site itself, but an understanding of the groundwater system in the surrounding area is equally necessary to assess the potential for groundwater pollution and the resulting consequences for local water supplies.

1

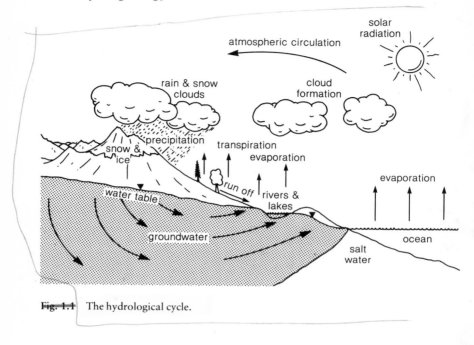

Fig. 1.1 The hydrological cycle.

1.2   The hydrological cycle

All hydrological and hydrogeological textbooks have an introductory chapter describing the hydrological cycle, and the reader should refer to some of the books listed in the bibliography in order to obtain a detailed understanding of the various processes which are involved. Groundwater forms part of this cycle and, in order to understand how a particular groundwater system works, it is necessary to examine the other elements in the cycle. This section contains a brief description of the hydrological cycle as a background to groundwater studies, which will both explain the need for

the various types of field measurements described in subsequent chapters, and serve as a check-list in planning your own field investigations.

The hydrological cycle is a vast and complex system which circulates water over the whole planet. Figure 1.1 illustrates the various parts of this cycle, which both starts and ends with the oceans. Energy from the sun powers the system, causing water to evaporate from the surface of the world's oceans which then vaporises to form large cloud masses. These clouds are moved by the global wind system and, when conditions are right, the water precipitates, falling back to the surface again as rain, snow

2

or hail. Some of the water falls on to the land and collects to form streams and rivers which eventually flow back into the sea, from where the process starts all over again. Not all rainfall contributes to the flow of streams and rivers in this way. Some of it is returned to the atmosphere as evaporation from water-bodies or the ground surface, and as transpiration from plants. A further portion of rainfall percolates through the soil to reach the water table and becomes groundwater. Groundwater usually flows through saturated rock under the influence of a hydraulic gradient which, in unconfined aquifers, is the water table. Rocks which both contain groundwater and allow water to flow through them in significant quantities are termed *aquifers*. Unless groundwater is removed by pumping from wells, it will flow through an aquifer towards natural discharge points, which comprise springs and seepages into streams or rivers, and also discharges directly into the sea. The property of an aquifer which allows fluids to flow through it is termed *permeability*, and this is controlled largely by geological factors. Properties of the fluid are also important, and water permeability is often called *hydraulic conductivity*. In both sedimentary rocks and unconsolidated sediments, groundwater is contained in and moves through the pore spaces between individual grains. Fissure systems in solid rocks can significantly increase the hydraulic conductivity of the rock mass. Indeed, in crystalline aquifers of all types, most

groundwater flow takes place through fissures and very little moves through the body of the rock itself.

Some geological materials do not transmit groundwater at significant rates, while others only permit small quantities to flow through them. Such materials are termed *aquicludes* and *aquitards* respectively, and although they do not transmit much water, they play a major role in controlling the movement of water through aquifers. Very few natural materials are completely uniform and most contain aquiclude and aquitard materials. Figure 1.2 shows how the presence of an aquiclude, such as clay, can give rise to springs and may support a perched water table above the main water table in an aquifer.

When an aquifer is overlain by impermeable rocks, the pressure of groundwater can be such that the level of water in wells would rise above the base of the overlying rock. In such instances the aquifer is said to be *confined*. Sometimes this pressure may be sufficiently great that the water will rise above the ground surface and flow from wells and boreholes without pumping. This condition is termed *artesian flow*, and both the aquifer and the wells which tap it are said to be *artesian*.

A groundwater system, therefore, consists of rainfall recharge percolating into the ground, reaching the water table, and flowing through rocks of varying permeabilities towards natural discharge points. The rate at which water flows through the system depends upon the rainfall,

3

(a)

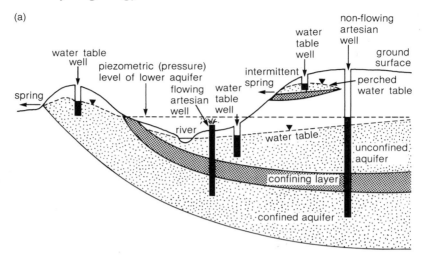

(b)

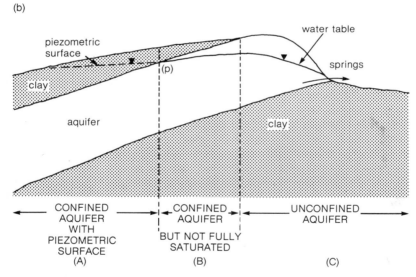

Fig. 1.2 The top diagram (a), shows two aquifers; a lower aquifer which is confined over much of the area; and an upper aquifer which is unconfined. The upper aquifer includes low-permeability material which supports a perched water table. The diagram shows the rest water levels in various wells in both aquifers. The lower diagram (b), shows

how both confined and unconfined conditions can occur in the same aquifer. In zone A, the aquifer is fully confined by the overlying clay and is fully saturated. The groundwater in this part of the aquifer is at a pressure controlled by the level of water at point (p), and water in wells would rise to this level which is above the top of the aquifer. In zone B, the aquifer is overlain by the clay but is not fully saturated and the groundwater pressure is the same as that in zone C. In zone B, the overlying clay will prevent any direct recharge. The aquifer in zone C is unconfined and is that part of the aquifer which receives direct recharge. Seasonal fluctuations in the water-table levels will alter the lateral extent of zone B along the edge of the aquifer. It is likely to be at a minimum at the end of the winter and at its greatest extent in early autumn, before winter recharge causes groundwater levels to rise.

evaporation, the geological conditions and many other factors. It is this system which the hydrogeologist is trying to understand by carrying out field measurements. Your hydrogeological investigation will define part of this overall cycle, as it exists in your area of study.

1.3   Stages of an investigation

John Barnes stated in *Basic Geological Mapping* (Open University Press) that 'much of the search for water is geological common sense'. He is quite correct, as no proper understanding of a groundwater system can be achieved without first understanding the local geology. A hydrogeological investigation, however, consists of more than geological mapping, interpretation and common sense. There are several phases to such studies, and the amount of effort put into each one, and the order in which they are carried out, is often varied to meet the needs of a particular set of circumstances.

1.3.1   Desk study

This is an essential prerequisite to any hydrogeological field investigation. It consists of the assembly of all available information, and provides an early opportunity to get a 'feel' for the groundwater system you are to study. It also enables the information which will be needed to complete this understanding to be identified, so that the collection of field measurements can be planned in detail. The desk study includes an examination of published geological information, so that potential aquifers can be identifed from their lithologies and their extent can be defined. An examination of topographic maps for the area will help identify spring lines, which may provide clues to the elevation of the water table if water-level measurements are not available. In Britain, along with many other developed countries, the geological and hydrogeological information required for a desk study is likely to be available. In other parts of the world these records may not be so easily acquired and this puts a greater emphasis on the fieldwork as a means to defining an area's hydrogeology.

5

1.3.2 Initial reconnaissance

Once a desk study has been completed, it is important to get to know the study area. In common with most other forms of geological fieldwork, this means putting your boots on and walking over at least some of the area. Where the investigation covers a large area it will not be possible to cover it on foot. Drive round as much of it as you can, and if possible stand on a hill or other vantage point so that you can see across the area. Picture the geology in your mind, and envisage how groundwater is flowing through the rocks, using information which you obtained in the desk study. It may be necessary to carry out additional geological mapping, and certainly you will need to look for such features as seepage and spring lines. It also gives an early opportunity to locate wells and to get to know the people who are using them. A very important part of this reconnaissance is to identify the type of information needed to complete your understanding of the groundwater system and which can only be obtained from the exploration phase of the investigation.

1.3.3 Further data gathering and evaluation

In many ways this part of the investigation overlaps with the reconnaissance study. Whenever a spring is located, record its position and measure its flow and possibly its conductivity and temperature. When-

ever you find a well, always measure and record its depth, the standing-water level and whether or not pumping was going on. In addition, try to obtain an idea of how frequently, and at what rate, the well is pumped. Any such extra information is then evaluated against that assembled at the desk study stage. An increasingly more detailed understanding of the groundwater system is built up in this way, and is used to plan the remaining stages of the investigation. The continual re-evaluation of data is an important feature of groundwater investigations, because the data available are usually sparse.

1.3.4 Monitoring programme

Once the initial reconnaissance of an area has been completed, it is usual to initiate a monitoring programme. It is important to measure and record, on a regular basis, the flow of springs and streams, the level of the water in local wells and how much water is being pumped from them. Rainfall and evaporation measurements may also be required. Such information will enable you to build up a picture of how much water is flowing into and out of the groundwater system. It is also important to take samples of water from springs and wells, to have them analysed and to examine their chemistry. Quite often, water chemistry is used to identify and 'fingerprint' groundwaters from different aquifers.

1.3.5   Exploration

It is very likely that the information you have been able to collect so far in the groundwater study is not enough to allow a proper understanding of the local hydrogeology. Additional information is often needed on the geology, the water levels in various horizons, or the hydraulic conductivity. This type of information can only be obtained by using such techniques as surface geophysics (see the companion Handbook, *Field Guide to Geophysics* by Milsom). This stage of the investigation may also include drilling exploration boreholes and conducting pumping tests on suitable existing or newly drilled boreholes, to assess the aquifer's hydraulic characteristics (see the companion Handbook, *A Field Guide to Water Wells and Boreholes* by Lewis Clark).

1.3.6   Water balance

Once the extent of an aquifer has been established and its boundaries identified, it should be possible to quantify the volumes of water which are passing through the groundwater system. The amount of recharge can be assessed using information about rainfall and evaporation. Discharges from the aquifer can be estimated from spring flow measurements, stream gauges and the amount of water pumped from local wells. This stage of the investigation constitutes a summary of all the previous work and is the point at which it becomes possible to start to answer those questions which caused you to initiate the investigation in the first place. These may include the availability of groundwater resources and the suitability of the resources for supply, the effects of new or increased abstractions, or the threat of pollution from a proposed waste disposal operation.

1.4   Hydrogeological report-writing

A great deal of information is drawn together during a hydrogeological investigation. This allows the groundwater system to be understood, and provides answers to specific questions asked by the person, company or authority who commissioned the work. At this stage, it is usual to write a report which describes the investigation and the conclusions which you have drawn. The introduction should state the purposes of the investigation, the terms of reference and the name of the body requesting the work. The body of the report will contain both information and interpretation, with the facts clearly separated from any inferences drawn from them. The final section should contain the recommendations, which are the actions which you think should be taken to fulfil the purposes of the investigation. If necessary, include information as to how the recommendations can be implemented.

The report should be written in a clear, straightforward way which the reader will be able to understand. The inclusion of carefully drawn maps and diagrams will help this understanding. It is good practice to present all the data collected in the field, usually in summary form in an appendix. In large reports it may be better to present these data in a separate volume. Useful advice on report-writing is contained in *Guidance Notes on Report Writing*, published by the Institution of Geologists (1985).

2
Instruments and equipment

The most frequent field measurement that a hydrogeologist is likely to make is the water level in a well or borehole. It is good practice to have a suitable 'dipper' handy in your car boot or the back of your truck. Tools will be needed to remove the cover from the top of the well, before taking a water level reading or a water sample. Special equipment is needed to take this sample, together with clean glass or polythene bottles. On-site readings may call for the use of a pH probe or at least some pH-sensitive paper, a thermometer and conductivity meter or other similar instruments. Flow measuring equipment could also be required, and this is described in Chapter 7. It may only be a kitchen measuring jug or a large bucket, but you will sometimes have to use more expensive equipment such as a current meter which measures stream flow velocities. In any event, a stop-watch will be needed for timing your measurement and on occasions, continuous recording equipment will also be required. You will always need a notebook to make a record of all field measurements and observations.

Some investigations will require the use of specialist equipment such as borehole geophysical logging probes to look at variations both in the rock and groundwater, or ion-specific meters for detailed groundwater chemistry studies. Some of this equipment is discussed in later chapters but generally instruction in its use is beyond the scope of this book. You should refer to the companion Handbook in this series: *Groundwater Quality and Chemistry*, by Mazor, or to one of the books listed in the bibliography, for further information on this equipment.

Finally, do not forget the 'geology' in hydrogeology. You are likely to need your rock hammer, perhaps a chisel, measuring tapes, map-case, hand lens and all the other paraphernalia found in a field geologist's knapsack. Table 2.1 lists this equipment and is based on information contained in Chapter 2 *Basic Geological Mapping* by John Barnes.

2.1 Water-level dippers

Among the hydrogeological fraternity, any instrument which is lowered into a borehole or well to measure the water level is called a 'dipper'. Despite

Field Hydrogeology

Table 2.1 Geological field equipment (after Barnes, 1981)

Hammer – ideally 1 kg weight
Chisels
Compass and clinometer
Handlens
Tape – 3 m 'roll-up' steel tape
Map-case
Field notebook with pencils and
  erasers
Scale rule, protractor and stereonet
Suitable field clothing

this being a jargon term, I intend to use it throughout this book. Over the past few years there has been an increased range of dippers brought out by different manufacturers. Most of these consist of a length of twin-core cable, which is graduated in metres, wrapped on a drum and has a pair of electrodes attached to the end. When the electrodes touch the water surface, a circuit is completed which activates either a light or a buzzer. Some dippers have ordinary round-section cable, with the depth graduations marked by adhesive bands. It is necessary to use a steel tape to measure the distance in centimetres from the nearest metre mark, to obtain a precise water level reading. These instruments are not so easy to use as the sort which has twin wires, one running down each side of a flat tape graduated in metres and centimetres. Figure 2.1 shows the main

(a) general assembly

(b) detail of probe

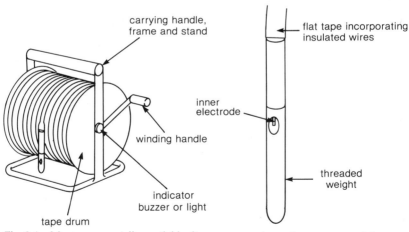

Fig. 2.1 Most commercially available dippers are made on the same general format. Batteries (usually totalling 9 V) are housed in the drum spindle, which also contains the electronic circuitry. The probe is usually made of stainless steel or brass and acts as an electrode, with a second inner electrode being visible through holes in the weighted end. This weight is usually threaded and can be unscrewed to give access to the inner electrode.

Fig. 2.2   The photograph shows several different types of commercial dippers currently in use in Britain. A total depth probe (2nd left) is included in the picture. These are used to measure the depths of wells and boreholes, and consist of a cylindrical weight suspended on 500 m of steel-wire cable. Depth markers are fastened to the cable at 5 m intervals. Artesian water levels are measured using the clear plastic tube at the front of the picture and the steel tape. The jug, bucket and stop-watch are used for spring-flow measurement.

features of such an instrument. This arrangement allows quick and easy readings to be taken, and the tape can be read to the nearest half centimetre if required. These dippers can be obtained in a wide range of sizes, from a compact ten metres up to an enormous 500-metre length (although it is difficult to imagine anyone needing to measure a water level at this depth!). When purchasing a dipper, ensure that it is long enough to measure the deepest water table in your area, bearing in mind that pumping-water levels in abstraction boreholes are much deeper than rest-water levels. A dipper of 80 metres or 100 metres length will cater for most situations, but 200 metres tape may be required for measuring some pumping levels in boreholes with large drawdowns. If you are studying a shallow aquifer

where there are only hand dug wells, a ten or thirty metre dipper should be adequate. It is important to make these considerations as the size of the dipper affects the cost and if you can get away with a smaller one they are very much more convenient to carry. Figure 2.2 shows several commercially available dippers.

2.1.1   Using a dipper

Water-level measurements should be taken by lowering the probe down the well or borehole until it hits the water, causing the buzzer to sound or the light to come on. When this happens, pull the tape back slowly out of the water until the signal stops. Repeat the exercise several times to enable you to 'feel' the water surface. It is conventional to take the point where

11

the buzzer stops or light goes off as being the water level. Use your fingers to mark the position on the tape against a fixed datum point such as the top of the casing, and then read off the level to the nearest centimetre. When using a dipper with a cable marked off in metres, use a steel tape to measure the distance from the nearest metre mark *below* your fingers and add the two values together. Record the dipped level and time of measurement and make a note of the datum used (see section 6.2). If the dipper is of the 'lamp type', position it so that bright sunlight will not prevent you from seeing when the indicator light comes on, perhaps testing the light before starting to make sure that no readings are missed. In some makes, the electrical circuit includes a mechanical relay which makes a 'click' as the light goes on and off. On the whole, fewer mistakes are made with audible dippers. It is a good idea to test the dipper each day before lowering it down a borehole. The usual way is to unscrew the weighted end of the probe to expose the inner electrode, and then to complete the circuit either by using the weight to short out the contacts or by immersing the end in water. The commonly used, unhygienic alternative is to wet the fingers with spittle to do the same job, but this is not good practice. If the lamp or buzzer does not work, check the batteries (and bulb), and should these be working, inspect the cable for breaks. The first thirty metres of cable gets most wear, and the insulation can

be rubbed off as the tape is lowered over sharp edges such as the top of the casing. Try to reduce this wear as much as possible by using softer materials such as wood or your hand. If the insulation is worn through to the wires, they may break, thereby preventing the circuit from being completed. Do not be tempted to cut off the damaged section and reconnect the probe, otherwise you will always have to remember to deduct a constant length from your readings and errors will inevitably occur. Occasionally, the circuitry inside the drum may be faulty and will need checking and repair by an electrician. It is worth remembering that dippers will not work in groundwater with a low conductivity. Such conditions are unusual and occur in aquifers with a low chemical reactivity, such as pyroclastic rocks. Great confusion can be caused in the field when the dipper works at the surface but fails down the borehole. In these circumstances try an improvised dipper which 'feels' the water surface (see below).

2.1.2 Improvised dippers

If a commercial dipper is not available, it is possible to make your own using the sort of electrical cable used for door bells, provided that it is weighted adequately to make it hang straight. Make sure that the two electrodes cannot touch accidentally and that the only way a circuit can be completed is when both are underwater (see Fig. 2.3). Before you use the probe in a well, test it in a bucket of

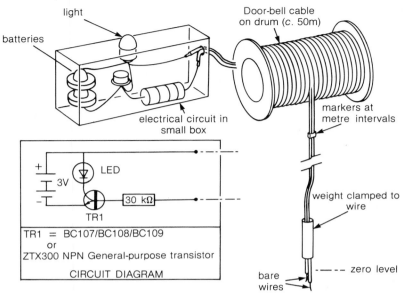

light
batteries
Door-bell cable
on drum (*c.* 50m)
markers at
metre intervals
electrical circuit in
small box
weight clamped to
wire
LED
+
3V
30 kΩ
−
TR1
zero level
TR1 = BC107/BC108/BC109
    or
ZTX300 NPN General-purpose transistor
CIRCUIT DIAGRAM
bare
wires

Fig. 2.3 This home-made dipper comprises a length of door-bell cable and a simple electronic circuit which is made from components which are available from any electronics supplier. The cable should be weighted to make it hang straight, and then hung to allow it to stretch before markers are fixed at metre intervals. Special adhesive numbers can be obtained, but care must be taken during use to ensure that they do not move or come off. The electronic circuit consists of a transistor, a 30 kΩ resistor, a LED display light and is powered by two 'button' batteries. Make sure that the two bare ends of the cable cannot touch and the circuit will then only be completed when both electrodes are submerged. Test the circuit in a bucket of water before using in the field.

water. Divide the cable into metre graduations, using electrical insulation tape as markers. This improvised dipper should be used in the same way as a commercial one, with a steel tape being used to measure from the nearest marker, as described above.

An old-fashioned way of measuring a water level is the 'wetted string' method. Tie a weight to a length of string and then rub the string with coloured chalk. Lower the weight into the well until it is submerged. Now pull it out and lay the string in a straight line on the ground. Measure the depth to the water level with a tape, using the point where the chalk was washed off to indicate the water surface. This method is cumbersome, slow to use and susceptible to errors if water is flowing into the well from levels higher than the standing water level. It can be improved a little, however, by using a surveyor's tape instead of string. The technique will

13

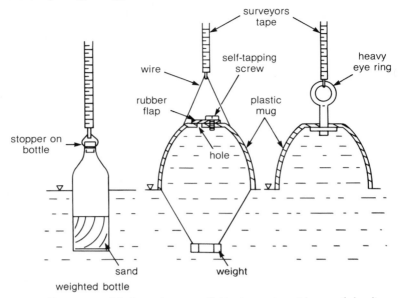

Fig. 2.4 If a commercial dipper is not available, improvise with one of the dippers shown in this diagram. In each case it is important to ensure that the 'instrument' hangs vertically, and to test it in a bucket of water before using it in a well. This will enable the difference between the water surface and the zero mark on the surveyor's tape to be measured. Add this value to all readings.

usually work with water levels to depths of 25 metres or so, but requires the operator to know the approximate depth beforehand to ensure that the end is submerged.

There are several other types of home-made probe which all work by 'feeling' the water surface by a change in the weight on the line and are generally more satisfactory than the 'wetted string' method. In each case either a length of string or a surveyor's tape is used to lower the device down the well. These devices are shown in Fig. 2.4. The first type uses a small screw-top bottle as a float. The sort of bottle used for soft drinks is very

suitable. Weight the bottle with sand so that it will float upright, screw the top on tightly and then tie the bottle securely to the measuring tape or string. The device needs to be calibrated before use, by being lowered into a bucket of water until the weight on the string reduces as the bottle starts to float. The distance between the water surface and the zero on the tape should then be measured. This value should be noted and added to all readings taken with that particular device.

Another type of home-made dipper is the 'plopper'. There are two main types of plopper, which are also shown in Fig. 2.4. Both are made from a plastic handleless mug (or even an old can), which is weighted so that it will sink when it hits the water. The mug is attached to the tape in such a way that when it is pulled back out of the water, the flooded cup is inverted and emerges bottom-end first. As it is pulled out of the water, atmospheric pressure will keep the cup full until the rim leaves the water surface. At this point, the water flows out of the cup and the weight on the line is dramatically reduced, enabling the water surface to be 'felt'. Each reading should be repeated two or three times before you note it down. Once again the device will have to be calibrated in a bucket of water to measure the constant value to add to all field measurements. These improvised methods will work best in large-diameter wells with a shallow water table. They are not recommended as a permanent substitute for commercially available dippers, but there may be times when no alternative is possible.

When making water-level measurements in wells or boreholes used for water supplies ensure that you adhere to the hygiene precautions given in section 10.7.

2.1.3 Measuring artesian heads

In boreholes where the water level is

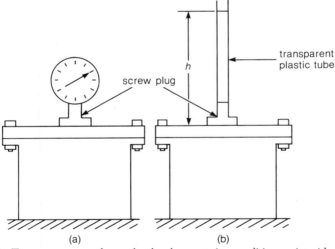

(a)   (b)

Fig. 2.5 To measure groundwater levels where artesian conditions exist, either use a pressure gauge which is calibrated in metres head of water (a), or fix a transparent plastic tube to the borehole (b) and measure the water level as a head (h) *above* the datum point. Choose a robust plastic tube and fix a metal screw fitting to one end. It is sometimes difficult to keep the tube absolutely vertical, but this will not matter, provided that you measure the *vertical* distance between the datum and water surface (see also Fig. 6.6).

above ground level, head measurements must be made in a different way. Ideally, the borehole should be capped with a blank flange which is fitted with a threaded dipping plug. The plug is removed and replaced with either a pressure gauge or a transparent tube (see Fig. 2.5). The pressure gauge gives an accuracy of ±0.5 metre or better, which is suitable for regional monitoring. It has the advantage of being quicker and simpler to use than the tube. A transparent tube gives an accuracy of ±1 cm and is better for pumping tests. If the heads are more than two metres above ground, practical considerations may dictate the use of a pressure gauge. A long tube can be used if it is fixed to a convenient tree or telegraph pole. Remember it does not matter what shape the tube takes, provided that the vertical distance above the datum is measured.

2.2 Water-level recorders

Where a continuous record of groundwater levels is required, a recorder of some sort must be used. The most usual type uses a float to sense the water level. This is connected to a chart on a revolving drum. There are two sorts of this float-operated recorder; each has a revolving drum which is covered with a paper chart, and a clockwork mechanism for the time. In the first type (Fig. 2.6) the float has a steel tape attached to it which passes over a pulley and has a counter-weight at-

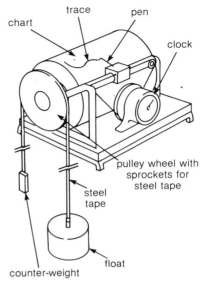

Fig. 2.6 Horizontal drum recorders are arranged so that the drum revolves in response to changes in water level, while the pen is driven by a clock. In operation, the instrument is enclosed by a metal cover and usually is further protected by a lockable box. Beware of condensation inside the box which may ruin the paper charts. Also look for insects, snakes and other animals taking up residence; some of them may have a poisonous bite.

tached to the other end. As water levels change, the float responds, moving the pulley which is attached through a system of gears to a horizontal chart drum. A pen moves horizontally across the chart, driven at a steady rate by the clockwork mechanism. Changes in water level are thus recorded by the pen's trace. The scale of the graph so produced can be changed by selecting the appropriate gears to alter both the

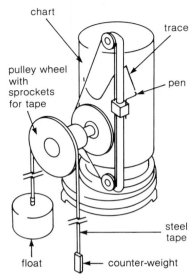

chart

trace

pulley wheel
with
sprockets
for tape

pen

steel
tape

float

counter-weight

Fig. 2.7 Vertical drum recorders are arranged so that the pen moves in response to water-level changes, while the drum is rotated by a clockwork mechanism.

time scale and the vertical scale as required.

The alternative type of arrangement has a vertical chart drum which is driven round by the clockwork mechanism. In this case, the pen is moved vertically by the rotation of a pulley wheel which is attached to the float in the same manner as before (see Fig. 2.7).

Horizontal drum recorders have the advantage that they will continue to function when there are large changes in water level, such as during a pumping test. The disadvantage is that the chart must be changed within the specified time period or the record will be lost as the pen jams against the end of the drum.

Vertical drum recorders will continue to function for as long as the clock continues to drive the chart drum, so records will not be lost if you are late in changing the chart. However, the range of water-level changes that they can accommodate is limited by the height of the drum, reducing their value in pumping tests where large changes are likely. This last disadvantage can be overcome by altering the gearing, but it is sometimes difficult to read off accurate times and levels from a small-scale trace. Try to avoid letting the charts over-run, as the later part of the trace is superimposed on the earlier section, which can lead to confusion.

In this electronic age, more sophisticated equipment is available for recording water levels using a pressure-sensing transducer. The transducer is installed in a borehole at sufficient depth below the water surface to ensure that it will not be exposed by changes in level. Fluctuations in the water level cause a corresponding change in the weight of water above the transducer. This varies the electrical resistance of the device, and when a small electrical current is passed through it, the variation in voltage is recorded at the surface. This information can be converted directly into water level data. Transducers can be attached to chart recorders, but they lend themselves to being used in conjunction with an electronic data-logger which records information in a form which can be read directly into a computer.

Whichever type of recorder you

17

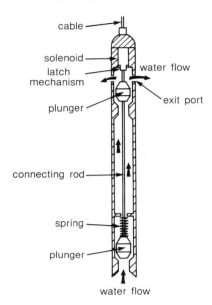

cable

solenoid

latch
mechanism

plunger

water flow

exit port

connecting rod

spring

plunger

water flow

Fig. 2.8 A sample of groundwater can be obtained from a predetermined depth using a depth sampler. Water continuously flows through the instrument as it is lowered down the borehole, so that when it stops it contains water from that particular depth. The plunger-like valves are closed either electrically (as in the diagram), or by dropping a weight attached to the cable. Electrically operated samplers are usually lowered on the cable which carries the electrical charge. Mechanical samplers are lowered on a small-diameter (2–4 mm) steel-wire cable. The valve mechanism is activated by a 'messenger' weight which is clamped to the cable and allowed to free-fall to the sampler. Beware of using this type of sampler at shallow depths, as without much water to slow it down, the messenger may knock the sampler off the cable.

use, however, it will be necessary to use a dipper to calibrate it, each time the site is visited either to change the chart or collect electronically recorded data from a logger.

2.3 Water samplers

It is often necessary to take a water sample from a well or borehole on which there is no pump. To do this you must lower a depth sampler to a predetermined level, activate it and obtain a large enough sample for chemical analysis. Figure 2.8 shows the workings of a typical depth sampler. It consists of a stainless steel tube, open at each end, which can be closed by spring-loaded bungs and has a capacity of 1 litre. The device is

lowered down a borehole to the required depth, when the spring mechanism is activated pulling the two bungs into place. This may be done mechanically by dropping a weight attached to the cable on which the sampler has been lowered. Alternatively, some makes are activated using a high voltage charge as a trigger. Care should be taken when sampling just below the water surface, if using the mechanical messenger type of sampler, as it is easy to knock the sampler off the end of the wire.

Another type of sampler is available which uses compressed air to control the depth at which water is drawn into the sampler, and an example is shown in Fig. 2.9. The instrument consists of a brass or stainless steel cylinder, some 15 to 20 mm diameter, fitted

Fig. 2.9 The photograph shows a selection of depth samplers and related field equipment. The two winches at the left are used with electrically operated samplers which are seen attached. The control box for these samplers is seen between the winches. The smallest winch is the pressure type and is charged with air using the foot-pump. A 500 ml glass sample bottle is shown together with a small plastic bottle used for a sample for toxic metals determination. The other small bottle is made of glass and is fitted with metal caps. It is used to hold samples for gas analysis. Several 500 ml bottles are in the carrying holder, with small bottles for 'metals' samples next to it on the right.

with a pressure-release valve at the base. The top of the cylinder is attached to a length of nylon tube wound on to a drum. This nylon tube is approximately 5 mm in diameter and can be anything up to 100 metres long. The system is pressurized using a foot pump or compressed air, or nitrogen from a gas bottle. A pressure gauge is attached to the drum and also to a separate reel of cord which is used to haul up the sampler when it is full. In operation, the system is pre-pressurized to a value in excess of the hydrostatic pressure at the depth from which the sample is required. When calculating this pressure, remember that 1 bar is approximately 10.2 m head of water. Once it has been pressurized, the cylinder is lowered to the selected depth and the air pressure is vented. The relief valve will now open, allowing water to flow into the cylinder. When a sufficient volume has been obtained, the system is pressurized again to close off the release valve. The sampler is then hauled to the surface using the nylon cord, and emptied by manual activation of the relief valve. The main advantage of this type of system is its slim size, which allows a sample to be

19

taken from small-diameter boreholes or where access is limited. The main disadvantages are the limited volume of sample (100–150 ml), which may mean several trips, and a somewhat cumbersome means of retrieval.

If a depth sampler is not available, you can improvise one, using a weighted bottle. Choose a one-litre size glass bottle, capable of withstanding high pressure, e.g. one which has held beer or lemonade. It needs to have a narrow neck, otherwise restrict the opening by punching a small hole (2–5 mm) in the cap which is left on. Make sure that the bottle is thoroughly clean, so as not to contaminate the sample.

Weight the bottle with a suitable object of at least half a kilogram and firmly secure strong string around the neck. Calculate the depth from which you want to take your sample; remember, it should be from below the stagnant water inside the casing. This information should be available from the driller's records or by talking to the owner. Mark this length on the string and then wind the string on to a stout stick. To take the sample, lower the bottle into the well as quickly as possible until the predetermined depth is reached. As the bottle sinks, the air trapped inside it will bubble out through the restricted opening and prevent a significant inrush of water before you reach the depth you want. Once you have let out all the string, wait for a couple of minutes or so, to allow all the air to bubble out. Then carefully haul the bottle to the surface.

When taking a depth sample, bear in mind that the water obtained is unlikely to be totally representative of groundwater at that depth in the aquifer. Groundwater flow in the borehole is common and usually affects the groundwater quality within the borehole by mixing or chemical reaction.

2.3.1 Sample bottles

It is essential to have a supply of clean glass bottles in which to put your samples for transport back to the laboratory. In most cases 500 ml is large enough, but some laboratories require 1-litre samples. Use a funnel to help avoid spillage, and always rinse the funnel and bottle with a little of the sample before you fill it to the top. It is usual to take a sample of about one litre, which provides enough water for rinsing out and also a separate 50 ml sample for heavy metal determinations. This small sample should be put into a separate bottle, usually of polythene, which contains 1 ml of nitric acid (HNO_3), thereby keeping the metals in solution.

Several parameters are often measured at the well-head, as well as in the laboratory. This is important in order to assess how much change has occurred between the sample being taken and the analysis being carried out. These changes are generally due to the reduction in pressure of the groundwater sample since it was removed from the bottle. A general description of the appearance of the sample should always be made, e.g. clear, cloudy, etc. The usual parameters measured in the field include pH, temperature and conductivity. Probes are available which will enable you to take these measurements very easily. Indicator paper can also be used to measure pH, although this does not give such accurate results.

Once you have filled the bottles, take great care not to drop them. Wet glass is slippery and you do not want to have to repeat depth samples if it can be avoided – they are hard work! Simple carriers are available (see Fig. 2.9), which are also useful in preventing samples from rolling about in your car-boot. Make sure that you label the bottles very clearly to identify the borehole or well and the depth from which the sample was taken. Table 2.2 lists the information

Table 2.2   Information to be recorded in the field for groundwater samples

(1)   General information
Date: day/month/year
Time: hours/minutes (indicate time zone, daylight saving, etc.)
Sampler: name and initials
Sample point: location where sample was taken, e.g. Blogg's No. 2 borehole
                   or Brown's spring
Description:
Sample point
Number: reference number in well catalogue, etc.
Total depth of borehole (in metres)
Rest/pumping water level (in metres below datum); state the pumping
    information and give details of datum
Sampler's comments: give any relevant information to help to interpret the
    data
General appearance of the sample – e.g. clear, cloudy, coloured (state which)

(2)   Field measurements
Depth sampled (give depth below datum and specify datum used)
Pumped sample (state how long the pump was running before the sample was
    taken)
Temperature (°C)
Dissolved oxygen (DO) (mg/litre)
DO % saturation
Alkalinity to pH 4.5 (mg/l as $CaCO_3$)
Eh (mV)
pH
Conductivity (μsiemen/cm measured at field temperature)

which should be recorded on the label. In addition, make a record of each sample in a field notebook. The same information should be written on the label of 'metals' samples. If you are taking samples in and around a waste disposal site, or from where the groundwater could be contaminated, wear rubber gloves to prevent getting any sample on your skin. Ensure that the sample cannot spill in your car, and avoid having samples which may contain volatile substances in an enclosed space such as a vehicle. Clean the sampler and other gear afterwards with detergent and rinse well with clean water. Do not use it in other boreholes, *especially ones used for a water supply.*

2.3.2 Pumped samples

Pumped samples are usually obtained from supply wells, or during pumping tests on new wells and investigation boreholes. They are not likely to show up variations in water chemistry at different levels within the aquifer, but will provide valuable information on the quality of water which is being used for supply. Samples taken over a period, the duration of a pumping test, for example, will show up any likely variations with time. Ideally, samples should be taken from a tapping on the pumping main, at a point as close to the well-head as possible. Try to avoid taking samples from a weir tank, but if this cannot be helped, make sure that a sample of water is taken from the discharge pipe and not out of the tank.

In some circumstances it is necessary to take a pumped sample from observation boreholes where you will have to install a temporary pump. For example, two or three cubic metres are required for some isotopic measurements, and in a study of dissolved gases, samples need to be pumped to ensure that pressures are maintained and de-gassing does not occur. This type of fieldwork would require special equipment such as a small submersible pump, generator, lifting tackle and a higher level of staffing than usual.

2.4 Ground-level measurements

In most groundwater investigations it is important to know the relative ground levels at the places where you take groundwater level and spring discharge measurements. This enables you to interpret flow directions and to determine the aquifer units being drained by each spring.

These levels can be estimated from topographic maps to an accuracy of less than half the contour interval. On British Ordnance Survey 1:25 000 scale maps this interval is either 25 feet or 10 metres, so you should expect to be able to estimate levels to ±5 m which may be adequate for a general idea of groundwater flow directions in hilly areas, but will not be good enough in most cases.

A better measure of altitude can be obtained using an altimeter, which is basically an aneroid barometer graduated in metres height. It is important to calibrate an altimeter before and after each time that it is used, by taking it to a place of known elevation. It is also important to know what changes have occurred in atmospheric pressure during the period the altimeter was in use. If large changes have taken place, then either the field values must be corrected, or abandoned and the exercise repeated. Information on short-term variations in atmospheric pressure is best obtained from a recording barograph. However, these are expensive instruments and usually it is not worthwhile to buy one (you would be better off investing in a hand level). In many countries, meteorological stations have a recording barograph. As these stations include fully instrumented government stations, airports and airfields and many high schools, there is an excellent chance that it will be possible to find one in the general location of your study area.

Robust, pocket-watch sized instruments are available, such as the *Thommen Mountain Altimeter* and these will give much more accurate level values than you can estimate from contour lines on a topographic map. In areas of low relief, however, where it can be expected that the gradient on the water table is small, this technique will not be sufficiently accurate, and should not be used. The likely errors of measurement will be greater than the differences in elevation of the observation points, which will almost certainly lead to significant mistakes in estimating the direction of groundwater flow.

Be careful if you are travelling by air, and take your altimeter in your hand luggage. Although aircraft cabins are pressurized, baggage compartments seldom are. Most altimeters only register up to 5 000 metres above sea-level, and may not function properly if they are subjected to altitudes of two or three times this maximum.

An alternative method of determining your elevation is described by Barnes in *Basic Geological Mapping*, and employs a hand level. This is an instrument which enables you to sight a truly horizontal line. Some geologists' compasses such as the Brunton and Meridian models have one built into them, while the Abney level is designed as a hand level. The method involves standing at the point where you require a level, and looking round the surrounding countryside with the hand level for a prominent feature at approximately your own elevation and where the ground level can be established accurately from information on your topographic map. Provided that you can find a feature no more than 1 km from your position, and within ½° of your level, it should be possible to determine your elevation within 10 m. Figure 2.10 illustrates this technique.

The most accurate ground-level measurements are obtained by using a surveyor's level. This type of instrument is essentially a telescope mounted on an adjustable base, which enables the line of sight through the telescope to be maintained in a horizontal position. Ground levels are

Fig. 2.10 Levelling in a contour by hand-level. Set the zero and then search for a feature within ½° of your level-line. (Reproduced by permission of J W Barnes.)

measured against this horizontal line, using a levelling staff – a sort of long telescopic ruler. Levelling using this type of equipment is a straight-forward procedure, and with care you will be able to obtain results to within a centimetre or two. Of course, professional surveyors or civil engineers who use this type of instrument frequently can expect an accuracy of a few millimetres.

Levelling is a relatively straight-forward technique, and a full description can be found in Bannister and Raymond (1984). The surveying techniques that you require are not difficult, can be easily learned, and are soon acquired with practice.

2.5 Soil and rock samples

A thorough knowledge of an area's geology is essential to understanding its hydrogeology. The basic equip-

ment of a field geologist will be needed from time to time, therefore, to enable you to examine such things as the lithologies of aquifers, to map joint and fissure systems, or otherwise supplement the available geological information. You may even have to carry out field mapping in some parts of the world. The basic equipment that you will need includes a note-book (with adequate pencils, pens and erasers), rock hammer, chisels, compass–clinometer, handlens (×10), steel tape, acid bottle, map-case, grain-size scale, sample bags, labels and a waterproof marker pen. You may find it useful to include a number of the Handbooks from this series in your rucksack – a full list of titles is given on page ii of this book.

2.5.1 Grain-size analysis

A rock's permeability can be inferred from its lithology, and this may be used as the basis of an initial assessment. A grain-size chart (see Appendix III) is a very useful and quick way of estimating particle size in both unconsolidated sediments and indurated sediments. For more accurate measurements with unconsolidated sediments, a set of sieves (see Fig. 2.11) is required. These are really laboratory equipment and are not easy to use in the field.

Laboratory sieving consists of drying and weighing the sample – which has been crushed gently to ensure that individual grains are separated. The sample is then passed

Fig. 2.11 The photograph shows several sieves of the type used in grain-size analysis. (Photograph provided by Endecotts Limited.)

through a series of up to 19 sieves, either by being shaken (dry sieving) or by being washed through (wet sieving). Both methods involve the use of a small brush, which ensures that all particles smaller than the sieve mesh will pass through it. It is important not to allow the finest material to be lost, particularly with wet sieving. The dry weight of the sample retained on each sieve is then measured, and the data from the test are then usually plotted as in Fig. 2.12.

Small pocket-size sieves are available which can be used in the field. Usually the sieve has several interchangeable, rimmed gauze discs, each with a different mesh size. Small quantities are brushed through each sieve in turn, and then weighed if possible. This method enables a more accurate estimate of the grain-size distribution to be made than is possible by using a grain-size chart such as the one in Appendix III. The results will only be an approximation nevertheless, and this method should not be used where the accuracy of laboratory methods is required.

The grain-size distribution of an unconsolidated material can be estimated using an improvised method, which is based on the principle that fine grains settle out of water more

Field Hydrogeology

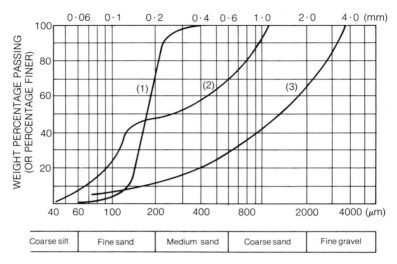

Fig. 2.12   After sieve analysis, a particle size distribution curve is constructed by plotting a graph on semi-logarithmic paper. The data are expressed as the percentage weight passing (or percentage finer than) the mesh size of each sieve. The particle sizes (in μm or mm) are plotted on the logarithmic scale, using the size ranges as shown in Appendix III. In the three example curves shown, (1) is a uniform sand, (2) is a poorly graded fine to medium–coarse sand and (3) is a well-graded silty sand and gravel.

slowly than coarse grains. All you need is a narrow, parallel-sided glass jar with a screw lid, such as a 450 gram (1 lb) jam jar, a grain-size chart (Appendix III), 10× handlens and a ruler. Half fill the jar with a representative sample, and then fill the jar with water, adding a few drops of water glass (sodium silicate, Na_2-$Si_4O_9.H_2O$) if available. This will act as a flocculation agent, helping the suspended grains to settle faster. Make sure that the sample is broken up, by stirring it into a slurry. After securing the lid, shake the jar vigorously to ensure that all the sample is in suspension, then stand it on a firm base for 24 hours to allow it to settle.

The sample should then be graded in a fining-upward sequence. Examine the sample through your handlens, using the grain-size chart to help distinguish between clay/silt, fine/medium sand and coarse sand. Take care when handling the jar, to avoid disturbing the sample. Measure the thickness of each layer and use this to estimate the proportion of each grain size in the sample. Figure 2.13 shows the equipment and illustrates the method with an example.

2.5.2   Sample bags

Polythene bags which have a snap fastening are available in a wide range

26

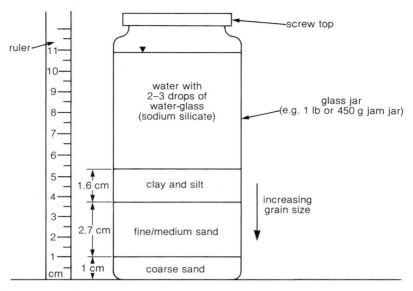

Fig. 2.13 Improvised grain-size analysis. Place the sample in a glass jar to no more than half-full. Add water to near the top with two or three drops of sodium silicate (water-glass) if available. Shake and stir thoroughly, and then stand for 24 hours or until the water is clear. Estimate the grain sizes from a chart (Appendix III) and calculate the proportions as in the example:

Clay and silt = 1.6/5.3 × 100 = 30%; fine/medium sand = 2.7/5.3 × 100 = 51%

coarse sand = 1.0/5.3 × 100 = 19%

of sizes. These are ideal for holding small samples for taking back to base. If samples are being taken for laboratory testing, you will need at least 1 kg each of clays, silts and sands; 5 kg for fine and medium gravel; and 30 kg for coarse gravel. The same size samples are needed for indurated rocks with similar lithologies. Use strong bags and see that they are properly sealed, double wrapping if necessary. Clearly label each sample with a reference number and the position and depth from where it was taken. Use self-

adhesive labels and a waterproof pen. Record details of each sample in a field notebook, using the reference numbers to avoid samples being confused. It is also a good idea to mark the position where you took each sample on a large-scale map.

2.6 Tool box

A wide assortment of tools is required for a variety of jobs which you are likely to face when working in the

Field Hydrogeology

Table 2.3 Tools and equipment check-list

Tool box
18″ adjustable pipe wrenches
Small spanners or socket set
4 lb hammer and claw hammer
Assorted chisels and crowbar
Manhole-lifting keys
Hacksaw, screwdrivers (large and small), pliers, allen keys
Woodsaw, knife, brace-and-bit
Steel tape, builder's spirit-level
Spade, bricklayer's trowel, bucket, hand-brush, wire brush
Electrician's tape, assorted pieces of electrical wire and cable
Assorted nails and screws, 'super' glue, bolts, nuts, washers, releasing agent

Other equipment
Dipper, with spare batteries, depth-sampler, bottles, funnel, pH-meter or
    indicator paper, specific ion probes, thermometer
Total depth probe, stop-watch, measuring-jug, pressure gauge, transparent
    tube
Torch (non-sparking type, in case of explosive gases)
Calculator, notebook, pencils, etc.

Safety equipment
Safety harness with rope, safety helmet, safety shoes/boots, high-visibility
    jacket, goggles, rubber gloves
Gas detector, miner's safety lamp, first-aid kit.

See section 10.2 on safety equipment.

field. A basic list is given in Table 2.3, and illustrated in Fig. 2.14. The list can be used to check off equipment, before setting out from base. This is always worth doing, because there is nothing worse than being halfway through a job, only to discover that an important tool has been left behind.

2.6.1 Removing well covers

Access to many wells is through a manhole cover. These are often made of cast iron and range widely in size and weight. Some have lifting handles incorporated into them, while others require the use of special lifting keys. Before removing a cover, clean off any soil, vegetation or other debris which may inhibit your lifting the cover easily, or which could fall into the well once the cover has been removed (see section 2.1.3). Use a spade and a hand-brush to clean the cover, and do not forget the dirt which has slipped down into the small gap between the cover and the frame. Take care when lifting the cover not to put either your hands or feet under it and risk fingers or toes being broken if the cover slips.

28

Fig. 2.14 This photograph shows the range of tools and equipment that a hydrogeologist is likely to need in order to take field measurements. Some basic items of safety equipment are also included. (See Table 2.3 for a basic list.)

It is tempting to use a chisel or the edge of the spade as a lever to prise up covers where the lifting handle is missing or when your lifting keys do not fit. This approach is not to be recommended; always take great care. Caution should also be exercised when attempting to lift heavy covers. Do not try this on your own; some of these covers are over 50 kg which could cause back sprains. Boreholes may have bolted flange covers fitted to the top of the casing (see Fig. 6.6). Long-handled adjustable spanners or pipe wrenches, and perhaps a releasing agent, may be needed to remove them. Sometimes the borehole is situated in a small chamber below ground level, which restricts access and prevents the use of long-handled span-

ners. A socket set will come in handy here. Try to avoid using the wrong tools; this will almost certainly mean that you will damage the bolt heads and you are also more likely to slip and hurt your hand.

2.6.2 Field equipment

A variety of tools are needed when installing field instruments or making running repairs in the field. These include woodworking tools, pliers, screwdrivers, a knife and assorted screws, bolts, nuts and washers. Builders' tools will be required when springs are being modified for measurements to be taken. See Table 2.3 for a more comprehensive list.

3
Sources of information

Information for a hydrogeological investigation is required on a wide variety of topics. Geological records in the form of published maps and reports and borehole records are needed to define the extent of aquifers, identify possible recharge areas and parts of aquifers which are potentially confined or unconfined. Hydrometric measurements are used to help quantify recharge and build up a picture of how much water is flowing through each aquifer. The extent of the main aquifers, water-table contours and the location of well, springs and other groundwater features are plotted on topographic base-maps. The amount of this type of information which you may find available will vary both from one country to another and within each country. In all cases you are likely to have to supplement existing information with your own field measurements. Appendix II provides more detailed advice and a list of organizations who may hold this information on a worldwide basis. Generally, in Western Europe, North America and Australasia there are government and local government organizations which collect and collate the sort of data that

you may use. Over much of the rest of the world the situation is very variable. In some countries topographic maps and remote-sensing data are restricted for security reasons and access is limited. Permission to view these records is often needed from high-level government or military authorities. In other countries, a large number of government departments may hold relevant information. In some parts of the world it may be difficult to obtain permission to use the data which exist and diplomatic skills may be as important to the hydrogeologist as his technical abilities. Do not be too disappointed if it proves impossible to obtain information. A list of recent projects is often a good way to begin to search for where information is held. Published papers may be useful occasionally, but they are frequently non-specific or out of date as far as data sources are concerned. In some countries, however, papers published in the 1950s or 1960s are still regarded as standard works. Sometimes the information may be free of charge, but in most cases it will have to be paid for, either bought as a book, report, or map, or as a fee to inspect or copy records.

3.1 Topographic maps

Maps will be needed to find your way around the study area, plan field trips and to record the position of features such as wells and springs as you find them in the field. All field geologists must be skilled in the use of maps and be able to locate their position accurately on maps of any scale. A great deal of useful information on position-finding on maps, and the use of geographical and metric grid co-ordinates is given in *Basic Geological Mapping* (Barnes, 1981).

Topographic maps of various scales are of value in hydrogeological studies; the scale used depends on the application and the available choice. Large-scale maps such as 1:25 000 or 1:10 000 are useful in the field because they allow accurate position-finding. Smaller scales such as 1:50 000 or 1:100 000 may be better suited for plotting field results or preparing maps to go into your report. In the latter case the deciding factor in choosing a suitable scale may be one which allows the whole area to be shown on a piece of A4-size paper. In practical terms the choice will be limited by the published maps which cover the study area. Give some thought to the scales where a choice is possible, but also think about the overall suitability of the map for your intended uses. For example, heavily coloured ornament may make it difficult to record information directly on to the map. Try to select a map which has topography shown by contours, so that the ground level of springs and

at well-heads can be estimated.

In some instances you may have to prepare your own map based on published material. Re-drawing will help you get rid of unwanted detail and produce a map which covers the study area. This will be necessary where the area of interest falls on the boundary of more than one map. Scales can be modified by using a photocopier which has enlargement and reduction facilities. It is important not to become confused over the scale changes during this process. The best way to avoid any problem is to ensure that a linear scale is drawn on the original map before any photo-reduction or enlargement is carried out. Otherwise, remember that if the paper size is doubled the linear scale is altered by a factor of $\sqrt{2}$. To change a scale from 1:50 000 to 1:25 000 photographically for example, the paper size must be doubled *twice*, i.e. A5-size paper must go through A4 to A3. The same rules apply when reducing scales. It is worthwhile remembering that enlarging a map does not make it more accurate, it just makes it bigger. Do not forget that the majority of published maps are copyright and that permission will be required from the copyright owner and perhaps a fee paid.

It is often necessary to measure the area on a map covered by a particular feature. Commonly used methods include counting grid-squares on the map or the use of a planimeter – an instrument designed to measure areas. If you do not have a planimeter, trace the area on to squared graph-paper and then count the number of squares, to obtain a value in square centimetres. Then apply the scale factor to convert the area measured on the map to the area on the ground. For example, an area of 7.62 cm² on a 1:50 000 scale map represents 1.91 km² on the ground. Look at the linear scale on the map to see how many centimetres represent one kilometre and then calculate the number of square centimetres on the map which represent one square kilometre, to arrive at the correct scale factor.

3.2 Geological information

The most usual form in which geological information exists is as a map. There is a great variation, however, in the scale and accuracy of published maps and the detail which they contain. The cover of geological mapping is incomplete in many countries. Where maps are available check the date when the survey was carried out, as generally the older the mapping, the greater the risk of outdated information, especially in matters of detail.

If it is not possible to buy a published map which covers your study area, check with local public libraries and museums to see if they have a copy of any out-of-print material. Water authorities or local government departments may also have copies that they will allow you to consult. Many geological maps have associated reports or memoirs; ask for these as well as the maps. Universities and colleges are another likely source

and where they include a geology department, they may possess unpublished material such as undergraduate mapping or postgraduate theses which cover your study area. Remember that the area may have been mapped by students from universities and colleges from far away. A discussion with local academics is likely to help you discover who may have undertaken such work.

Published books and scientific papers are a further source of information. These range from publications by the geological survey or other government departments, books published by scientific publishers, to scientific papers published in journals which may have an international circulation or merely a local one. The United States Geological Survey has undertaken a number of groundwater investigations in countries throughout the world and these reports are a valuable source of geological and groundwater data. Information on published reports can be obtained from the USGS in Denver, Colorado (see Appendix II). A common way for a list of publications to be compiled is to examine the references listed in the publications that you find in local libraries. It may also be worthwhile contacting local geological societies or even those who are interested in a much broader range of natural history. Quite often one or more of the members will be a professional or even an amateur geologist with a particular interest in the geology of your study area. The local libraries or museums should be able to provide information and the names and addresses of such societies. In Britain many such societies and scientific publishers are listed in *The Geologist's Directory*, The Institution of Geologists, 1985.

3.2.1 *Unpublished records*

In addition to the types of records listed above, a great deal of geological information is contained in borehole records which include borings for water, stratigraphic correlation and mineral and energy resources exploration. Such information is often collected by geological surveys and government departments concerned with the development of natural resources. Mining, quarrying and oil companies also have their records but may not be willing to give you access to them.

Shallow boreholes are often drilled as part of site investigations for new civil-engineering works. Usually, records of such boreholes are kept only by the organization who drilled them, but it may be worthwhile contacting local highway, sewerage and water authorities or other organizations involved in large-scale construction work in the study area. Once again, even if such records exist, you may not be allowed access to them.

3.3 Groundwater levels

In some countries groundwater level measurements are made on observation boreholes, in addition to those on abstraction boreholes. Water-supply authorities and companies, as well as geological surveys and other government organizations, may collect such records. It is important to know how these groundwater level measurements have been taken and whether they are pumping or rest levels (see section 6.3). It is unlikely that sufficient groundwater level measurements will be available for you to properly define the groundwater flow regime in the study area, and extra field measurements may be required.

3.4 Surface water measurements

Routine flow measurements are taken on many major rivers in most countries. Although the majority of attention is given to the major rivers, smaller rivers and streams may have short-term records or a few 'spot' readings where instantaneous measurements have been taken by current-meter methods. It is usual for such measurements to be taken by government departments concerned with resources, agriculture, irrigation or power; or local government bodies responsible for water supply, flood alleviation or water-resources management. Records may be expressed as average daily flows, total daily flows or instantaneous flows. It is important to ascertain which applies to the records that you are examining (see Fig. 3.1).

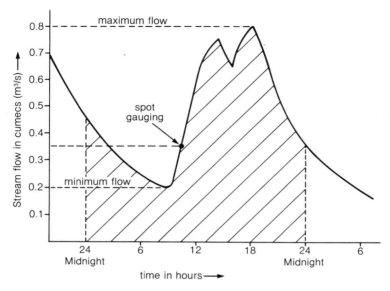

Fig. 3.1 Confusion may arise from the way in which flows are recorded and the units used. The graph represents the flow of a stream as measured continuously at a gauging station. During the 24-hour day represented by the shaded section, the *minimum flow* was 0.2 cumecs and the *maximum flow* was 0.8 cumecs. The *total daily flow* is represented by the shaded area on the graph and is 0.48 cumec-days. This is easier to understand if it is converted to 41 472 m³/d by multiplying by the number of seconds in a day. This figure can be used to calculate the *average daily flow*, which is the arithmetic mean. The value in this instance is 0.48 cumecs or 41 472 m³/d. During the late morning, a spot gauging was taken by current-meter (see section 7.4) and this produced an *instantaneous flow* value of 0.35 cumecs.

3.5 Rainfall and evaporation

Rainfall measurements are made in most countries by a government department or central government agency. Additional measurements may be made by water authorities, catchment authorities, university and college departments and schools. Similar organizations also take other meteorological measurements, including open water evaporation and

other parameters needed for calculating evaporation losses.

3.6 Abstraction records

This information is often collected at the same time as a well inventory is prepared because a record of abstracted quantities is often only available from industrial abstractors. In some countries, abstraction information is collected by government departments and agencies. This may be in monthly or even annual totals. Daily abstraction rates may vary over a wide range at different times of the year and this fact may be masked by the way that the records are kept.

Where abstractors are not required to keep a record, few do so of their own volition. In such circumstances the quantities abstracted can be estimated from the pump capacity and the number of hours run. Alternatively, the use to which the water is put can be used as a basis for the estimate (see Table 9.1).

3.7 Groundwater chemistry

Information on the chemistry of water pumped from wells may be available from the abstractors or the organizations which collect groundwater level records. Collate as many chemical analyses as you can. Groundwater

quality can change with time, and isolated values may not be representative.

3.8 Air photographs and remote sensing

Aerial photographs are taken from aircraft for many purposes, and the use of satellite and air-borne imagery has grown in recent years. The availability of aerial photographs varies greatly from country to country. In many instances such data are obtained primarily for military purposes and their use is very restricted. Satellite information is more readily available from the Landsat satellite operated by the United States government, or from Spot Image operated by the French government.

Records are usually kept by central government departments or agencies and some local authorities.

4
Desk study

Once information on an area has begun to be accumulated, it is possible to begin a desk study. This is the examination of the hydrogeology of the area using all the available information gleaned from sources such as those listed in the previous chapter, and initially, without additional information being obtained from the field. It is usually carried out in an office (hence the name) and at first glance it may seem strange to include a description of such an exercise in a practical field guide. However, by defining what is already known about an area and identifying which questions need answering, a desk study can be a very useful way of planning a fieldwork programme. An early appreciation of the different types of field measurements which are needed will enable equipment to be assembled and plans to be made, so that the fieldwork can be completed over the minimum period possible. For example, some information may only be available during a particular season. It is important therefore, to plan so that the necessary field measurements are taken at the earliest opportunity, thereby avoiding delay.

A desk study also provides an early assessment (or model) of the groundwater system in the study area, which can be used as a model against which to compare field results. Such comparisons give confidence in the field activities and provide a means for modifying field measurement programmes to ensure that adequate information is obtained in the most efficient manner.

It is important to use a systematic approach for the desk study, gradually examining all the available information. The procedures outlined in this chapter will in some cases require specialized analytical techniques to interpret the available data. A description of these methods is to be found in later chapters, where the relevant field measurements are described. Where little information is available on an area, the desk study will be completed quickly and without producing much idea of how the local groundwater system works. The desk study should be completed nevertheless, as it plays an essential part in planning the field investigations. In extreme cases, a desk study may not be possible without preparing maps of the study area, using aerial photographs or satellite imagery, carrying out geological mapping to identify the aquifers and undertaking all the hydrometric field measurements: from rainfall to river flows.

Field data should be regularly incorporated into the desk study model as they are collected, so that the fieldwork programme can be modified when necessary. Changes in the frequency of field measurements, the inclusion of additional wells or springs in a monitoring network, or the termination of other measurements, are some of the modifications which are likely to be made during the course of an investigation.

4.1   Defining the area

The size of the area to be examined depends on the reason for the study. If the hydrogeology is being assessed to determine the likely effect of a new borehole abstraction on surrounding supplies, then an area with a radius of two kilometres will often suffice. For abstractions of less than 20 m^3/d, one kilometre radius is normally adequate, except in some fissured formations where a larger area may be needed. If the work is being carried out as part of a waste disposal investigation, a similar sized area to an abstraction study (2 km radius) should be examined. Never regard the boundaries of the study area as rigid. Once the survey is complete, it may be a good idea for any wells or springs which lie only a little more than two kilometres from the new borehole to be included in the study.

If the work is being carried out to assess the overall groundwater resources of an aquifer, then a much larger area must be covered to include the whole aquifer and extend a little way on to neighbouring strata. Attention must be paid to stream catchment boundaries, which are fixed by the topography and often do not coincide with groundwater or aquifer boundaries.

This type of investigation can cover areas varying from less than one km^2 in the case of a minor aquifer such as a river terrace, to several thousand km^2 for major aquifers. Once the size of the study area has been worked out, base-maps can be prepared as described in Chapter 3.

4.2   Identifying the aquifers

At the same time that you are considering the extent of your study area, you must decide which materials are likely to behave as aquifers and which are not. Follow the techniques described in Chapter 5, using all the available geological information to classify the rocks in the area as good aquifers, poor aquifers or non-aquifers. It may help to sort this out if you prepare a simplified version of the geological map using these categories.

The next step is to examine the aquifer boundaries which lie within the study area. It is possible for

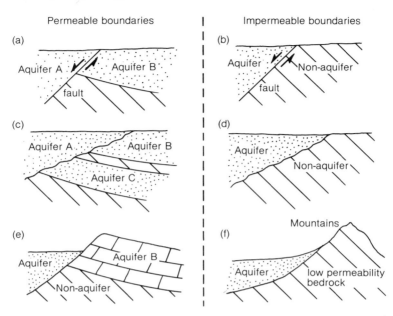

Fig. 4.1 Several different types of aquifer boundaries are shown in this diagram. In (a), two aquifers are faulted against each other. The boundary may be permeable and allow flow across it, with hydraulic conductivity being locally increased if rock fracturing has occurred but it could be impermeable. This can usually be detected only by differences in groundwater chemistry or from pumping-test analysis. The faulted boundary shown in (b) is essentially impermeable, as the aquifer is faulted against an aquiclude. In (c), aquifer A unconformably overlies a sequence which includes two separate aquifers. Conditions are similar to example (a), except that groundwater quality may be different in all three aquifers. Groundwater conditions in aquifer (d) are similar to those in (b) in that the aquifer overlies a non-aquifer. In example (e), an unconsolidated aquifer, such as a river terrace, partly overlies a lithified aquifer (such as limestone or sandstone) which forms an escarpment. Groundwater flow is possible between the two aquifers and, because of the topographic differences, it is most likely to be from aquifer B to aquifer A. In these circumstances, scarp-slope springs may not be seen. In example (f), a gravel aquifer overlies low-permeability bedrock. No groundwater flow takes place across the boundary, but the aquifer recharge will be greatly enhanced by surface runoff from the mountain area.

groundwater to flow across these boundaries? There are many sorts of boundary, but in broad terms these can be simplified into permeable and impermeable. Figure 4.1 shows a few examples of each type. It is important to examine this question, as it has a significant bearing on groundwater recharge and flow directions, and there are implications for groundwater quality and the pumping characteristics of wells.

Next consider the thickness of the aquifer and how it varies across the study area. This has a significant affect on its transmissivity. Borehole information will be particularly useful in this exercise. Lithological information will help define the base of the aquifer. Some formations, such as indurated mudstones, igneous and metamorphic rocks, are capable of acting as aquifers only where they are fractured or weathered. Therefore, in such materials it is important to examine borehole logs for information on the depth of weathering. In high latitudes perennially frozen ground, called permafrost, has a significant effect on aquifer characteristics and groundwater conditions. Frozen ground has a much reduced permeability compared with the unfrozen material, with the differences often being several orders of magnitude. Consequently, permafrost acts as an aquiclude or an aquitard, reducing recharge and confining the groundwater in the deeper unfrozen aquifer. Do not neglect the upper boundary of the aquifer. Is it at outcrop or is it overlain by other deposits? Outcrop areas are potential recharge areas, but any overlying low-permeability rocks may confine the aquifer, thereby having a fundamental influence on groundwater conditions.

4.3 Groundwater levels

Once the main aquifers have been identified and their boundaries defined, the groundwater which they contain can be considered. At this stage of the investigation it is unlikely that many direct water-level readings are available, but it is possible to interpret information from the topographic map using basic principles, and to obtain a general idea of groundwater flow directions.

Topographic maps can be used to locate possible spring lines by using the 'springs' and the starting points of streams which are shown on the maps. Poorly drained areas are usually underlain by low permeability materials and may mean that groundwater is discharging from an adjacent aquifer. Comparison of such poorly drained areas with the geological map will help identify aquifer discharge areas and boggy areas caused by underlying low-permeability materials.

Groundwater flows from areas of recharge to areas of discharge, with the general shape of the water table usually being a subdued version of the surface topography. Hence, once discharge areas have been identified, it is possible to infer the general direction in which the groundwater is flowing, from the local topography. This will provide a clue as to where the recharge areas lie. Other indications include a general lack of streams and other surface water features. Significant natural recharge can only occur at outcrop or through permeable superficial deposits. Therefore a further consideration of the geological and topographic maps will

Field Hydrogeology

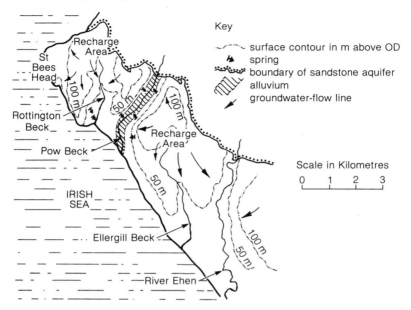

Key

~⌢~ surface contour in m above OD
🔥 spring
🌊 boundary of sandstone aquifer
〰 alluvium
↙ groundwater-flow line

Scale in Kilometres
0   1   2   3

Fig. 4.2 Groundwater flow map for part of the St Bees Sandstone (Permian) aquifer in West Cumbria, England. This map was drawn at the early part of the desk-study stage, before any boreholes had been drilled to provide an early idea of groundwater levels. See Fig. 6.18 (North West Water data).

enable likely recharge areas to be defined and groundwater flow directions to be deduced. A word of caution must be included at this point. Beware of having too much confidence in your deductions at this stage. Treat them as a hypothesis which must be tested using field evidence. Be particularly careful in fissured aquifers, especially karstic limestones, as groundwater can flow in different directions from those indicated by signs at the surface.

Figure 4.2 shows an example of how groundwater flow directions were deduced at the desk study stage of an investigation of the St Bees Sandstone aquifer in West Cumbria, England. Compare this map with Fig. 6.18 to see how the desk study

information was either confirmed or modified by field information.

At this stage, consideration should be given to the field measurements which are required to complete the hydrological study. This study is likely to include spring flow monitoring and groundwater level measurements, but before this can be planned in detail it will be necessary to complete a survey of wells and springs, as described in Chapter 6.

40

4.4 Surface water

By the time you reach this part of the desk study you will have already given the surface water system some consideration. In many areas, rivers and lakes receive at least some, or even all of their water from the ground (see Fig. 1.1) and thus surface water flow records are needed, so that the groundwater discharges from the aquifer can be quantified. Those streams and rivers which derive water from an aquifer in the study area should be identified, and any flow records should be examined using the method outlined in section 7.7. Once the available information has been compiled, it should be possible to decide if additional flow measurements are needed. Sometimes it will be adequate to measure the flow upstream and downstream of the study area to see whether flows increase. Where large-scale pumping tests on new boreholes are involved, however, it is likely that continuous records will be required to see if the new pumping reduces stream flows. This is only likely to be important where the pumping rate is equivalent to fifty per cent or more of the dry-weather flow of local streams.

4.5 Recharge

In order to estimate recharge to an aquifer, information is needed on rainfall and evaporation. Assess whether or not existing rain gauges and the nearest meteorological station

where evaporation is measured, are close enough to the study area to provide accurate information. If there are not enough, new ones will have to be installed. Chapter 7 provides information on both rainfall and evaporation measurements.

To decide whether there are adequate rain gauges, compare the annual average values for all rain gauges in the vicinity and their altitude. Areas of higher altitude generally have a higher rainfall. If the rain gauges cover a similar altitude range to the study area and have similar average values when any differences of height are taken into account, then it is possible to avoid establishing any new gauges. If the groundwater investigation is likely to go on for any significant period, then it will probably be worthwhile to set up at least one new station.

The best method for calculating the rainfall over a particular area, based on a number of rain gauges, was developed by Theissen. Draw lines between each gauge and bisect each line with a perpendicular. Then extend each perpendicular to form a series of polygons as shown in Fig. 4.3. It is then assumed that the rainfall for the area defined by each polygon is equal to that of the rain gauge at its centre. This method is much more accurate than a simple arithmetic mean of all the rain gauges in the area.

Owing to the difficulties involved in the evaporation measurement, it is best to use information from an established meteorological station if pos-

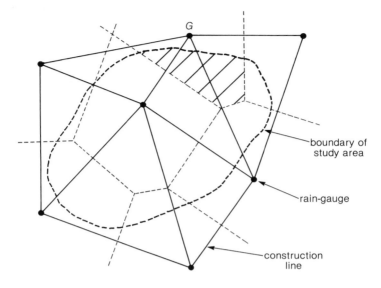

Fig. 4.3 Theissen polygons. Draw perpendicular lines to bisect the lines joining each rain-gauge, and extend them to form polygons. Catchment rainfall is calculated by assuming that the rainfall in any particular polygon is equal to that measured by the rain-gauge at its centre. For example the volume of rain falling on the shaded portion of the catchment is taken to be the depth of rain measured at rain-gauge 'G', multiplied by the area shaded.

sible. Remember that these values are potential evaporation, i.e. the maximum which can occur. Actual values of evaporation might be much less.

Provided that average values are available for rainfall and evaporation, it is possible to make an initial calculation of annual recharge. The value obtained will be somewhat crude, but an early estimate is likely to be useful where a new groundwater develop-

ment is proposed. If the estimated recharge is several times larger than the quantity of water required for supply, then the scheme is likely to succeed, but schemes which need all the available recharge are unlikely to be successful.

To make the calculation, subtract the evaporation value from the average annual rainfall to obtain an idea of how much rainfall is available for recharge (i.e. potential recharge). Then multiply the potential recharge value (in mm) by the area of the exposed aquifer. It is assumed that all the potential recharge will percolate into outcrop areas but only a proportion of potential recharge will soak into the aquifer where it is overlain by low-permeability material. It should be supposed that no recharge will take place through indurated mudrocks or

1	Average annual rainfall	987 mm
2	Average annual potential evaporation	450 mm
3	Potential recharge = 987–450	= 537 mm
4	Area of aquifer	
	(a) Outcrop	15.7 km²
	(b) Overlain by clayey drift	36.3 km²
	(c) Confined by mudstones	not used in calculation
	(d) Outcrop with urban area	2.5 km²
	(e) Clayey drift with urban area	3.2 km²
5	Calculation of recharge	
	(a) Outcrop: 15.7 × 537	= 8431
	(b) Clayey drift: 36.3 × 537 × 0.5	= 9747
	(c) Confined: nil	= nil
	(d) Urban area: 2.5 × 537 × 0.75	= 1007
	(e) Urban/clay: 3.2 × 537 × 0.75 × 0.5	= 644
		Total = 19 829 × 10³ m³/year

Fig. 4.4 This is an example of how a preliminary calculation is made of the recharge to an aquifer. The annual average quantity of rainfall which is available to recharge the aquifer is calculated as in the first three steps. The aquifer is then examined, and the areal extent of various types of surface conditions are measured. This information is then put together in (5) to calculate the annual recharge, which in this example is estimated to be approximately 19 800 × 10³ m³/year or 54 × 10³ m³/day.

where the aquifer is confined by clays. If the aquifer is not confined (see Fig. 1.2b), assume that half the potential recharge will reach either the aquifer by percolating through the clay or by running off the edge of the clay and then recharging the aquifer through the outcrop. In urban areas, recharge is reduced because rainfall runs off buildings and roads to sewers. In most cases this reduction is not likely to be by more than twenty-five per cent. An example is given of this type of calculation in Fig. 4.4.

4.6 Groundwater use

The availability of groundwater abstraction records is extremely variable. At best you are likely to find that information is only available for the largest abstractors and an on-the-ground survey will be necessary to locate the rest. Information on these larger abstractors will be useful to give an early idea of borehole yields in the area, and will indicate if there are any existing abstractors who may be affected by a new borehole source or threatened by a new waste disposal site.

4.7 Groundwater chemistry

Except for public water-supply sources, it is unusual for much ground-

43

Table 4.1 Chemical parameters to determine potability of a water supply (based on EC Directive 80/778/EEC)

Parameter	Unit	Guide level	Maximum acceptable concentration
Conductivity	μs/cm @ 20°C	400	1500
Chloride	mg/l Cl	25	400
Sulphate	mg/l SO_4	25	250
Nitrate	mg/l NO_3	25	50
Magnesium	mg/l Mg	30	50
Sodium	mg/l Na	20	175
Potassium	mg/l K	10	12
Calcium	mg/l Ca	100	250
Iron	μg/l Fe	50	200

water chemistry information to be available in existing records. There is often enough information however, to enable the potability of the water to be estimated. Table 4.1 lists the parameters which will help you decide on potability. Sometimes information is tucked away in the abstraction records for a borehole, or in a notebook where water levels are logged. The only way to find them is to painstakingly go through all the record books which are available. Modern analyses give the results in mg/litre, which is the same as the now old-fashioned 'parts per million' (ppm). Very old records may quote values in 'grains per gallon' and must be converted to mg/l by multiplying by 14.25 (assuming imperial gallons).

Be on the alert for potential problems such as saline intrusion, high nitrates from agricultural fertilizers, leachate from waste disposal sites, sewage effluent, especially septic tanks and cesspits, and runoff from mineral workings tailing dumps. All of these could cause serious problems for both new groundwater developments and existing supplies. The total concentration of dissolved minerals provides an idea of how long the water has been in residence in the aquifer; generally speaking, the longer the period, the greater the overall chemical concentrations. Groundwater pollution will complicate this picture.

It is possible to identify recharge areas using this principle, if sufficient data are available. Plot the distribution of values for total dissolved minerals or conductivity, on a map. Concentrations of low values are likely to be the recharge areas, with groundwater flowing in the direction of increasing values. Highly mineralized groundwaters are often an indication that the rate of groundwater flow is very slow, from which it follows that permeabilities are low or there are no natural discharge points. Examination of the geological structure will help you decide if groundwater is trapped in this way.

4.8 Aerial photographs

Aerial photographs are of great potential value in many aspects of field geology, and hydrogeology is no exception. Barnes (1981) gives a good description of the uses of aerial photographs for field geologists, and Way (1973) provides a handbook style treatment of aerial photograph interpretation methods, using interpreted photographs to illustrate these techniques.

In areas where there are no maps, a mosaic of aerial photographs can be built up and used instead. There are problems because the scale is only true in the centre of the photograph, and it will be increasingly distorted towards the edges; however, it is better than having no map at all.

Aerial photographs are particularly useful in a desk study, because they are very detailed and also show up features which cannot be seen easily on the ground. They can be used to find seepage areas, but the techniques of interpretation available to the hydrogeologist do not provide direct information about groundwater conditions. The general approach is to use the photographs to prepare maps showing variations in vegetation type, landforms, land use, soils and drainage. These maps are then used to interpret likely groundwater conditions, from which it is possible to define the best areas for new wells. Way (1973) explains many of the techniques used in this type of interpretation.

Infra-red aerial photographs are now quite common, and are especially useful to hydrogeologists because they are sensitive to temperature variations. As groundwater temperatures remain constant throughout the year, seasonal or diurnal contrasts with surface waters or the surrounding rock has proved to be an excellent method of locating springs and hence discharge areas. Satellite photographs are also proving their worth in hydrogeology. Unfortunately the scales of photographs which are commercially available limit their value in detailed studies, but they can be employed successfully in regional reconnaissance work.

Satellite photographs are produced by a method which reduces each picture to digitized data in computer format. This means that the data can be processed to give prominence to particular features such as temperature, and are very much more sensitive than the old-fashioned infra-red film. Similar techniques are now being employed from aircraft, using this line-scanning method. These surveys are very expensive, however, and are only likely to be used in large-scale hydrogeological investigations.

45

Table 4.2   Check-list for planning a fieldwork programme

---

1   Topographic information
Are adequate maps available? If not, use aerial photographs to produce a mosaic to use as a substitute base-map. Supplement with levelling where necessary.

2   Geological information
Is the available information adequate to define aquifer boundaries? Is additional geological mapping necessary? Are boreholes needed to provide geological information?

3   Groundwater levels
Carry out an on-the-ground survey to locate and record the position of all springs, wells and boreholes. Do aerial photographs show seepage areas? Can you draw reliable groundwater-level contours with the available data, or are extra boreholes needed? Decide on the need for a monitoring programme and details of frequency of observations, equipment required, etc.

4   Surface water measurements
Are extra flow measurements required? If there are, decide on suitable gauging sites and methods, frequency of measurements and the equipment that you will need.

5   Rainfall and evaporation
Are there adequate rain-gauges in the area? Where is the nearest meteorological station which provides evaporation figures? Do you need to take your own measurements? If you do, then decide on suitable sites for your instruments.

6   Groundwater use
Remember to include information on the volume and rate of water abstraction as part of the well-location survey.

7   Groundwater chemistry
Are extra samples needed? If yes, incorporate sampling into other fieldwork programmes.

4.9 Planning a fieldwork programme

As has already been stressed, planning the programme of fieldwork is one of the most important objectives of a desk study. By now you will have examined all the available information and identified what extra information is needed. An early objective of the fieldwork programme must be to obtain a first-hand knowledge of the area. If it is small enough, familiarize yourself with the area by walking over it when carrying out surveys to examine the geology or searching for wells and springs. If you have a very large area to investigate, it will be necessary to use a vehicle or perhaps even a helicopter, but try to walk over as much as possible. There is no substitute for personal observation to obtain a detailed knowledge of your area or an understanding of how the groundwater system works. By this direct experience, hydrogeological knowledge is built up which will enable you to interpret geological information in groundwater terms. This is the best way to become a competent professional hydrogeologist. A check-list for planning fieldwork programmes is contained in Table 4.2.

5
Field evaluation of aquifers

This chapter is concerned with the assessment of how water flows through an aquifer. Geological features such as lithology, petrology and structure, largely control the ease with which water will flow through the ground. Hence, by looking at these features, it is possible to identify the various flow mechanisms which exist in an aquifer. Besides examining these mechanisms of flow, it is important to consider the rate at which water is able to flow through the rock, as this governs the yield of wells and dictates the rate at which excavations will flood. It is also useful to know how much water is stored in an aquifer and can be extracted from wells. The characteristics which control groundwater flow and storage are usually referred to as the hydraulic properties. These can be measured in the field or laboratory but can also be assessed in general terms by consideration of the overall aquifer geology.

5.1 Hydraulic properties of aquifers

Groundwater flows through an aquifer when the water levels within it

are at different elevations. This difference in level is called *head loss* and is usually expressed in metres. The slope of the water table is called the *hydraulic gradient*, and is a dimensionless ratio of head to distance (Fig. 5.1). The equation which relates the velocity of groundwater flow (v) to the hydraulic gradient h/l is known as Darcy's law and has the following form:

$$v = k \times h/l$$

In this equation, k is the *hydraulic conductivity*, which is defined as the volume of water that will flow through a unit cross-sectional area of aquifer in unit time, under a unit hydraulic gradient and at a specified temperature. The usual units of hydraulic conductivity used by hydrogeologists are metres per day. (This is a reduced dimension from $m^3/d/m^2$.) Hydraulic conductivity is sometimes expressed in m/s and there are other units which are included in Appendix I.

Hydraulic conductivity depends on the properties of the aquifer to allow water to flow through it, and also on the density and viscosity of the water. These properties of water are affected

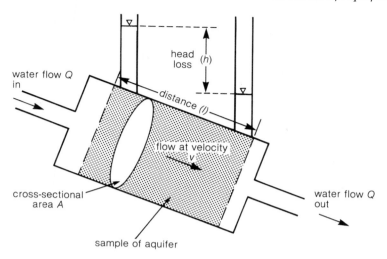

Fig. 5.1 Darcy showed that the velocity (*v*) of water flowing through a porous medium is equal to the hydraulic gradient (h/l), times a constant (*k*), which he called permeability. For a porous medium, the value of permeability varies according to the fluid involved, and water permeability is called *hydraulic conductivity*. As the amount of flow (*Q*) is determined by the velocity (*v*), and the cross-sectional area of the sample (*A*), Darcy's law can be used to calculate discharge.

by a number of conditions, such as the concentration of dissolved minerals, but the most important factor is temperature, because it alters the viscosity of the water. An increase in water temperature from 5°C to about 30°C, for example, will double the hydraulic conductivity. According to Darcy's law this will then double the velocity at which groundwater flows. As the temperature of groundwater generally remains constant throughout the year, this is not normally a serious problem for hydrogeologists, except in some shallow aquifers in areas of climatic extremes or in particular situations involving waste

water and industrial effluent. It can prove a problem where tests are being carried out, either in the field or the laboratory, to measure the hydraulic conductivity – if they involve pumping water into a test section or sample. It is important to ensure that the temperature of the test water is the same as the usual groundwater temperatures in that aquifer.

The property of a rock which controls the hydraulic conductivity is its *intrinsic permeability*. As this is a property of the rock, it remains constant whatever the fluids are flowing through it and applies equally well to oil and gas as to water. Intrinsic

49

permeability can be calculated when fluid density and viscosity are known. It has the reduced dimension of m^2 but is usually expressed in darcys.

It is usually more practical to measure groundwater flow in terms of the volume of water flowing through an aquifer, rather than its velocity. Darcy's law can be written as follows:

$$Q = A \times k \times h/l$$

where Q is the volume of water flowing in unit time through a cross-section with area A. Q represents a discharge and is measured as volume per unit time (for example m^3/d).

The amount of water which a rock can hold depends upon its *porosity*. This is the proportion of the volume of rock which consists of pores, and is usually expressed as a percentage. The principal factors which control porosity are grain size and shape, the degree of sorting, the extent of chemical cementation and the amount of fracturing. Figure 5.2 illustrates how porosity varies with grain shape and the degree of sorting in unconsolidated sediments. Those sediments which have been ideally sorted and have rounded grains of uniform size are the most porous. Porosity decreases as the angularity of the grains increases, because the grains pack together more closely. Similarly, as the degree of sorting is reduced smaller grains fill the pore spaces between the larger grains and porosity is also reduced.

In consolidated rocks, porosity tends to be lower than with unconsolidated sediments, because part of

(a) high porosity — rounded grains, uniform size (good sorting)

(b) low porosity — rounded grains, many sizes (poor sorting)

(c) medium porosity — angular grains, uniform size (good sorting)

(d) very low porosity — angular grains, many sizes (poor sorting)

Fig. 5.2 Porosity in unconsolidated sediments varies with the degree of sorting and with the shape of the grains. (Reproduced from Cargo and Mallory, 1974.)

the pore space is taken up with cement. Some rocks with relatively high porosity values may be poor transmitters of water because the individual pores are not interconnected. Figure 5.3 illustrates some of the aspects of porosity development in consolidated rocks. Porosity which has developed after the rocks have formed is termed *secondary porosity* to distinguish it from intergranular or primary porosity. Secondary porosity typically has two causes. Fracture porosity is caused by cracks in the rock associated with joints, bedding-plane fissures, tectonic joints and faulting (although where fault gouge

(a) vesicular
    porosity —
    may not be
    interconnected,
    e.g. basalt

(b) solution
    porosity —
    mild solution
    along crystal
    boundaries,
    e.g. limestone

(c) porosity along
    fractures or
    bedding planes

Fig. 5.3 Porosity in consolidated rocks. Note that example (c) shows bedding planes rather than individual grains, so covers a much larger area of rock. (Reproduced from Cargo and Mallory, 1974.)

has been produced or mineralization has occurred along the fault plane, groundwater movement will be restricted rather than enhanced). Secondary porosity is also caused by solution, which is common in limestones and other soluble rocks. Dolomitization also increases porosity because the magnesium ion is smaller than the calcium which it replaces. This process can produce an increase in porosity by as much as 13% of the total rock volume. The crystals are usually very small, so the increase in hydraulic conductivity is not as great. Porosity does not provide a direct measure of the amount of water that will drain out of the aquifer. This is

because a proportion of the water will remain in the rock, retained around the individual grains by surface-tension forces. This water is called the *specific retention*. The volume of water which will drain from the aquifer is termed the *specific yield*, and is a measure of how much water can be withdrawn from an aquifer under the influence of gravity.

5.2 Hydraulic properties and rock types

From the above discussion of the hydraulic properties of aquifers, it is obvious that a great deal can be learned about groundwater flow from the study of aquifer geology.

5.2.1 *Porosity and specific yield*

The proportion of porosity which is made up of specific yield is controlled by the grain size in non-indurated materials. This relationship is shown in Fig. 5.4. Specific retention decreases rapidly with increasing grain size, until it remains roughly around 6–8% for coarse sands and larger sized sediments. Specific yield is at a maximum in medium grained sands, because porosity decreases with increasing grain size. Note that a high degree of sorting will significantly reduce specific retention in coarse-grained sediments.

Although Fig. 5.4 is a best-fit curve based on scattered data, it can be used as a means of estimating likely values,

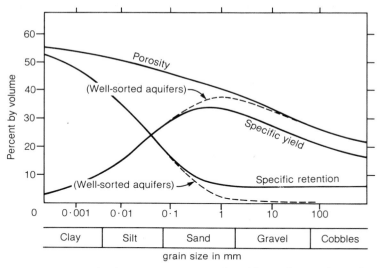

Fig. 5.4 The relationship between porosity, specific yield, specific retention and grain size for unconsolidated sediments only. The lines on this graph are best-fit curves drawn through scattered points and you should not ascribe any degree of precision to them.

Particle size	Assumed proportion of sample (%)	Typical specific yield (%)	Estimated specific yield (%)
Clay/silt	10	10	1.0
Fine sand	15	25	3.75
Medium/coarse sand	45	30	13.5
Fine/medium gravel	30	25	7.5
TOTAL			25.75

From total specific yield estimated as 20%

Fig. 5.5 In this example, the specific yield of an unconsolidated aquifer has been estimated using the grain-size distribution from a sieve analysis and estimated typical values for specific yield taken from Fig. 5.4. Each particle size is assumed to contribute to the overall specific yield of the aquifer in the same proportion as its volume. The components are summed and the estimated specific yield rounded down from this total. It must be stressed that this is a first-order estimate only, and the results should be treated with caution.

if you have obtained information about the grain size distribution for the aquifer. A method is outlined in Fig. 5.5. It is very important to remember that this is only a very rough estimate of the specific yield, and the results should be used with caution. Both porosity and specific yield can be measured in the laboratory but require adequate samples of aquifer material.

Table 5.1 shows the range of values of porosity associated with different unconsolidated sediments and solid rocks. Table 5.2 gives typical values of specific yield in a similar way. By comparing the two tables, it is apparent that the relationship between porosity and specific yield is more complicated in solid rocks than in unconsolidated sediments. From this comparison it is possible to get an idea of the effect of cementation and compaction in reducing specific yield and the influence of fracturing in increasing specific yield.

5.2.2 Permeability

The permeability of a rock is affected by the same geological factors as is its porosity. It is important, however, not to confuse porosity and permeability. Porosity is a measure of how much water the rock contains, whereas permeability determines how fast the water can flow through it. In this way, permeability and specific yield are

Table 5.1 Indicative values of porosity for a range of geological materials. Compare with Table 5.2

Material	Porosity (per cent)	Material	Porosity (per cent)
Coarse gravel	28	Loess	49
Medium gravel	32	Peat	92
Fine gravel	34	Schist	38
Coarse sand	39	Siltstone	35
Medium sand	39	Claystone	43
Fine sand	43	Shale	6
Silt	46	Till – mainly sand	31
Fine-grained sandstone	33	Till – mainly silt	34
Clay	42	Tuff	41
Medium-grained sandstone	37	Basalt	17
Limestone	30	Gabbro (weathered)	43
Dolomite	26	Granite (weathered)	45
Dune sand	45		

(Adapted from Water Supply Paper 1839-D by permission of the United States Geological Survey).

Table 5.2 Indicative values of specific yield for a range of geological materials

Material	Specific yield per cent
Coarse gravel	23
Medium gravel	24
Fine gravel	25
Coarse sand	27
Medium sand	28
Fine sand	23
Silt	8
Clay	3
Fine-grained sandstone	21
Medium-grained sandstone	27
Limestone	14
Dune sand	38
Loess	18
Peat	44
Schist	26
Siltstone	12
Till (mainly silt)	6
Till (mainly sand)	16
Till (mainly gravel)	16
Tuff	21

(Adapted from Water Supply Paper 1662-D by permission of the United States Geological Survey).

broadly related, so that in general, aquifers which have a high specific yield tend to be more permeable, and less permeable rocks usually have a lower specific yield.

In general terms, unconsolidated sediments tend to be significantly more permeable than their consolidated counterparts. This is because cementation has both reduced the overall void space in the rock and has

made the interconnection between pore spaces more restricted. As with porosity, the permeability of consolidated rocks (igneous and metamorphic as well as sedimentary) will be increased by jointing and fissuring. This is termed *secondary permeability*. Rock types can be classified on the basis of having primary permeability, secondary permeability, or both, and this has been done in Table 5.3.

Table 5.4 lists porosities and hydraulic conductivities (permeability in respect of water, remember) for a selection of unconsolidated sediments and rocks. Similar information on hydraulic conductivities is present in a slightly different form in Fig. 5.6. This information can be used to estimate likely values for the aquifers in your study area, in a similar manner to estimating specific yield but, in this case, read the hydraulic conductivity value straight off the scale in Fig. 5.6. Again, the same words of caution must be repeated. Remember that this will only provide a rough estimate of the hydraulic conductivity and that you must use it with care.

One aspect of permeability (or hydraulic conductivity) which should be borne in mind is the wide range of possible values it can have. The values of hydraulic conductivity given in Fig. 5.6 are on a logarithmic scale, hence the hydraulic conductivity of clean gravel can be expected to be around one thousand million (10^9) times greater than that of massive clay (i.e. nine orders of magnitude). In fact, it is quite common for hydraulic

Table 5.3 Classification of rock types in terms of permeability

Type of permeability	Sedimentary		Igneous and metamorphic	Volcanic	
	Unconsolidated	*Consolidated*		*Consolidated*	*Unconsolidated*
Intergranular	Gravelly sand, clayey sand, sandy clay		Weathered zone of granite-gneiss	Weathered zone of basalt	Volcanic ejecta, blocks, and fragments; Ash
Intergranular and secondary		Breccia, conglomerate, sandstone, slate; Zoogenic limestone, oolitic limestone, calcareous grit		Volcanic tuff, volcanic breccia, pumice	
Secondary		Limestone, dolomite, dolomitic limestone	Granite, gneiss, gabbro, quartzite, diorite, schist, mica-schist	Basalt, andesite, rhyolite	

Major rock types which behave as aquifers have been classified on the basis of the type of permeability which they exhibit. Intergranular or primary permeability is a feature of unconsolidated deposits and weathered rocks. It also occurs in most sedimentary rocks and those igneous rocks which have a high porosity. Secondary permeability is largely due to fissuring or solution weathering and only affects indurated rocks. (Adapted from *Groundwater in the Western Hemisphere* by permission of the United Nations.)

Table 5.4 List of indicative porosities and hydraulic conductivities for unconsolidated sediments and rocks

Geological material	Grain size (mm)	Porosity (per cent)	Hydraulic conductivity, K (metres per day)
Unconsolidated sediments			
Clay	0.000 5–0.002	45–60	$< 10^{-2}$
Silt	0.002–0.06	40–50	10^{-2}–1
Alluvial sands	0.06–2	30–40	1–500
Alluvial gravels	2–64	25–35	500–10 000
Consolidated sedimentary rocks			
Shale	Small	5–15	5×10^{-8}–5×10^{-6}
Sandstone	Medium	5–30	10^{-4}–10 (secondary permeability)
Limestone	Variable	0.1–30 (secondary porosity)	10^{-5}–10 (secondary permeability)
Igneous and metamorphic rocks			
Basalt	Small	0.001–1 (up to 50 if vesicular)	0.000 3–3 (secondary permeability)
Granite	Large	0.000 1–1 (up to 10 if fractured)	0.000 3–0.03 (secondary permeability)
Slate	Small	0.001–1	10^{-8}–10^{-5}
Schist	Medium	0.001–1	10^{-7}–10^{-4}

(Reproduced from S248 by permission of the Open University.)

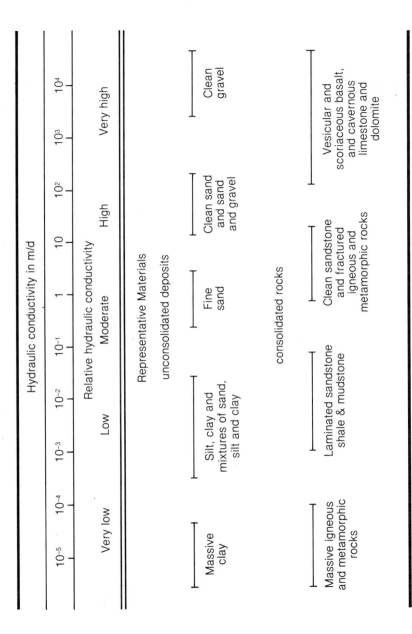

Fig. 5.6 Hydraulic conductivities in metres/day for various rock types. (Adapted from the *Groundwater Manual* by permission of the United States Department of the Interior.)

conductivities to be quoted to the nearest order of magnitude (e.g. 10^{-1}), rather than as a precise value. This reflects the reality of spatial variation within an aquifer due to geological factors, which basically means that it is virtually impossible to measure it accurately for an aquifer as a whole.

5.2.3 Dispersion

Another aspect of groundwater flow which is related to both the type of porosity and to the hydraulic conductivity, is the way in which a body of water is dispersed as it flows through the aquifer. It is easier to think of this body of water as a tracer (see section 7.7), which is injected into an aquifer as a fixed volume at a specific point. The tracer will not retain its original volume, because molecular diffusion and mechanical dispersion (see Fig. 5.7) will cause it to be diluted. This property of aquifers is important in tracer work and in the

consideration of the movement of pollutants through an aquifer. Attempts have been made to derive general relationships for the dispersion characteristics of aquifers, in a similar way to how hydraulic conductivity can be quantified. It has been found, however, that minor variations of both intergranular and fissure porosity, and also of permeability within an aquifer, can have very significant effects on the way in which dispersion occurs in any particular part of the aquifer. Dispersion is often assessed in the field, using tracer techniques (see section 7.9).

5.3 Assessing hydraulic properties

The relationship between hydraulic properties and geology can be used to classify the different rocks in the study area into potentially good aquifers, poor aquifers, and non-aquifers. You

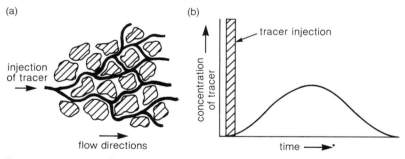

Fig. 5.7 As a tracer flows through a rock (a), it splits up each time an alternative pathway is reached. The result is that the tracer concentration is diluted by mechanical dispersion. Graph (b) shows how the concentration of a tracer varies with time as it flows past a particular point in an aquifer or emerges at a discharge point.

should be able to estimate likely values of hydraulic conductivity and specific yield for each formation, and decide whether the permeability is essentially primary, secondary or a mixture of both. Even in areas where good geological maps are available you are unlikely to be able to do this without going into the field, unless you already have a good knowledge of local lithologies and structure. This exercise should be carried out as part of your initial walk-over survey.

Look at the grain sizes and degree of sorting of sediments and sedimentary rocks. Use the grain size chart in Appendix III and carefully record the location where you take each reading. If necessary, bring samples back with you and have them tested or test them yourself (see Chapter 2). Inspect exposures of solid rocks for joints and other fissures and, if possible, examine cores recovered from boreholes in the area, to see if the rocks change with depth. It is important to remember that in most instances the size of individual fissures seen in rock faces is greater than that which occurs in the main body of the rock. This is because stress release takes place when the rock face is exposed by quarrying or erosion processes, and this allows fissures to widen. In aquifers where fissure flow is likely to be important, record the trend of joints and identify the direction of joint sets, plotting them on a stereonet if necessary. This will help determine the possible directions of groundwater flow. Your geological training should be adequate to enable you to complete these field exercises without undue difficulty. Where extra advice is needed, you will find it useful to have a copy of the appropriate Handbook in your rucksack (see the list in the front of this book). As information is gathered about the aquifers, draw it together in a summary form (see Fig. 5.8). You will now be able to use the estimated values of the aquifer properties and other information to help interpret groundwater level data and construct groundwater level contour maps and flow nets.

5.4 Using hydraulic properties information

An important reason for assessing the hydraulic properties of aquifers is to identify those formations which are likely to give the highest yields, thereby helping site new wells. Conversely, you may be looking for a suitable location for a waste disposal site and need an area with a very low hydraulic conductivity and no secondary permeability. The information contained in a summary such as the example given in Fig. 5.8, will enable you to decide which areas are worth further, more detailed consideration. In this example, any of the three main aquifers would ensure high yielding wells, but the success of a well in the Carboniferous Limestone would depend on sufficient fissures being encountered by the well, because all the permeability is through the fissure system. Likely sites for waste disposal may be found on either

	Grain size, sorting, etc.	Estimated hydraulic conductivity m/d	Estimated specific yield	Type of permeability	Notes
Main aquifer					
1. Glacial sands and gravels	medium/course sands and fine gravel with some cobbles	$10–10^2$	25%	primary	grain-size analysis
2. Triassic sandstone	fine/medium sets well cemented in parts	1–10	15%	primary + bedding fissures	confined by till in part
3. Carboniferous limestone	massive, dense rock	10^2	15%	secondary via fissures	some evidence of karst development
Poor aquifers					
1. Alluvium	mainly silt and thin sands	10^{3}	5%	intergranular	
2. Granite	weathering c. 1–2 m deep	10	5%	secondary via fissures	joint sets mapped
Non-aquifers					
1. Glacial clay	mainly clay, some silt	10^{5}	< 5%	primary	till and varved clays
2. Carboniferous mudstone	clay/silts	10^{4}	< 1%	secondary in weathered rock	

Fig. 5.8 Build up a picture of the aquifers in your study area, and their hydraulic properties, from field observations. Use the information contained in this chapter to estimate likely values of aquifer properties from which you can identify the main aquifers. Supplement this information with notes on relevant points. These notes are used in conjunction with a geological map of the area and groundwater-level information to complete the picture of the groundwater-flow system.

the glacial clay or the Carboniferous mudstone. It would be necessary, however, to ensure that the secondary permeability developed in the mud-stones by weathering did not allow polluted water to escape from the waste.

To assess the overall flow through

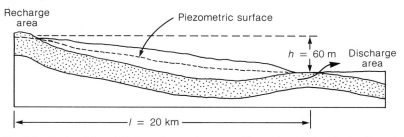

Fig. 5.9 Regional flow (Q) through a sandstone aquifer can be calculated using Darcy's law. The sandstone has an average thickness of 200 m and is 10 km wide. The distance from the recharge area to the discharge points is 20 km, and the head difference is 60 m on average. The hydraulic conductivity is 5 m/d. Substituting these values into Darcy's law we get:

$$Q = k\,Ah/l = 5 \times (200 \times 10\,000) \times 60/20\,000$$
$$= 30\,000\ \text{m}^3/\text{d}$$

If the specific yield is 15%, then the total volume of usable storage in the aquifer will be:

$$200 \times 10\,000 \times 20\,000 \times 0.15 = 6 \times 10^9\ \text{m}^3$$

an aquifer once a value for the hydraulic conductivity has been obtained, simply apply Darcy's law as shown in Fig. 5.9. It is also possible to work out how long it will take for water to flow through the aquifer, again using Darcy's law. Using the information from Fig. 5.9 and the first equation given in section 5.1,

$$v = kh/l = \frac{5 \times 60}{20\,000} = 0.015\ \text{m/d}$$

This means that it would take over 3 560 years for water to travel from the recharge area to emerge from springs in the discharge zone.

The usable volume of water stored in the aquifer can be calculated using the specific yield. This figure does not necessarily represent all the water which could be removed from the aquifer by pumping, because of practical limitations, but it is the upper

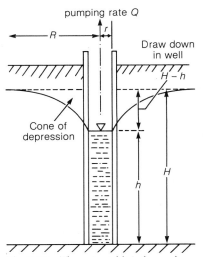

Fig. 5.10 The water table is drawn down into a cone of depression around a pumped well. Q, R, r, H and h are used in the equilibrium well equation (see text) for the determination of the hydraulic conductivity of the aquifer.

61

limit. When a well is pumped, the water level around the well falls in response to the pumping, forming a *cone of depression* (Fig. 5.10). The shape and extent of the cone of depression depends on the rate of pumping, the length of time pumping has continued, and the hydraulic characteristics of the aquifer. The amount that the water table has been lowered is called the drawdown (see Fig. 5.10). One of the many formulae which relate these parameters is the equilibrium well equation:

$$Q = \frac{k(H^2 - h^2)}{C \log(R/r)}$$

Q is the pumping rate; k is the hydraulic conductivity; H is the thickness of saturated aquifer penetrated by the well; h is the height of water in the well; r is the radius of the well; and R is the radius of the cone of depression. C is a constant with a value of 0.733; Q is in m^3/d, k is m/d and the other dimensions are in metres.

This equation can be used to estimate the yield of a well with different amounts of drawdown once they have stabilized, if the hydraulic conductivity is known, or it can be used to calculate the hydraulic conductivity if the drawdown and pumping rates are measured once groundwater levels have ceased to change.

5.5 Field measurement of hydraulic properties

Hydraulic conductivity and specific yield can be determined from a range

of field tests. All involve inducing a flow of water through the aquifer by pumping, and then measuring the change in water levels which result – both in the test well and a series of observation wells. Such a pumping test provides information which can be analysed mathematically using a wide variety of equations, such as the equilibrium well equation given in section 5.4. Other methods involve the analysis of the rate of change in water levels (i.e. non-equilibrium conditions). These techniques are used more frequently than the equilibrium formula, as it often takes a long time for steady-state conditions to be reached. The hydraulic properties calculated by these methods are a slight variation on the ones we have used up to now. Permeability is usually expressed in terms of *transmissivity*, which is the hydraulic conductivity multiplied by the full thickness of saturated aquifer. The volume of water contained in the rock is calculated as the *storage coefficient* (sometimes called *storativity*), which is exactly the same as specific yield in water table aquifers. The storage coefficient is defined as the volume of water released from a unit volume of the aquifer for a unit decline in head. When water is pumped from a confined aquifer, it causes a lowering of water levels, but this represents a reduction in pressure and not any dewatering. It is analogous to letting air out of a car tyre: where a measurable volume of air is removed but the tyre remains full, albeit at a reduced pressure. The storage coefficients of

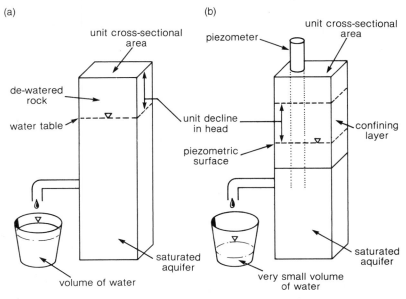

Fig. 5.11 This is a diagrammatic representation of the concept of the storage coefficient. In case (a), a water-table aquifer, unit decline in head produces a volume of water equivalent to the specific yield by dewatering a unit volume of rock. With a confined aquifer (b), the unit decline in head releases a relatively small drop of water and the aquifer remains fully saturated.

confined aquifers fall in the range 0.000 05 to 0.005, indicating that substantial changes in pressure are required over large areas to produce significant yields of water. This difference between confined and water-table aquifers is illustrated in Fig. 5.11.

All hydrogeologists at some time or another will be involved in carrying out a pumping test to determine the hydraulic properties of an aquifer. Pumping tests should never be carried out until you have obtained a general understanding of the groundwater

flow system in the aquifers in your study area. Without this knowledge it is impossible to design an appropriate test programme and monitoring network.

Pumping tests consist of controlled pumping from a test well with careful, detailed monitoring of groundwater levels in surrounding wells. These may be purpose-drilled observation boreholes or wells used for supply. Pumping tests can provide field data which enable aquifer transmissivity and storativity values to be calculated, provided that the correct analytical method is used. All such methods involve assumptions about the aquifer and it is vital, therefore, to choose the appropriate one. Pumping-test data can be used to select the correct submersible pump for a borehole to ensure optimum performance. They are also used to determine whether or not a new abstraction will affect local water supplies. In this case, measurements of springs and stream flows will also be needed.

Because of the complexities involved, a detailed description of pumping-test organization and analysis is beyond the scope of this basic field manual. When you are ready to tackle a pumping test, there are a number of useful references that you can use in the field. These include *A Field Guide to Water Wells and Boreholes* by Lewis Clark (1988), British Standard 6316 (1983), *Ground Water and Wells* (Driscoll, 1986) and *Pumping Tests* (Wright, 1985). Methods of analysing the data are contained in Kruseman and De Ridder (1983) and Rushton and Renshaw (1979).

6

Groundwater levels

A good set of reliable groundwater level measurements is the best foundation on which to build an understanding of a groundwater system. It enables you to define aquifer units, flow directions, changes in groundwater storage and overall aquifer discharge. A record of groundwater levels is also required so that the effects of pumping from a new borehole can be ascertained. Generally, two types of data are needed. To examine flow directions and the interrelationship between various aquifer units, you need to have as many observation points as possible, all read at approximately the same time. To consider changes in storage and examine the effects of pumping, long term records are required at a relatively small number of carefully selected sites. Most of these readings are taken at wells and boreholes, but this can be supplemented by information on the position and elevation of springs. All of these measurements depend upon a knowledge of the study area and details of all the wells, boreholes and springs within it. Hence, it is of fundamental importance in a groundwater study to compile a 'well catalogue', which

includes details of the depth and casing of each well in the area of interest.

In the first chapter, water table (i.e. unconfined) and confined aquifers were discussed. A record of groundwater levels will help you to decide whether a particular aquifer is confined or not. The simplest situation is a shallow well in an unconfined aquifer (Fig. 6.1a). It is likely to be lined with unmortared bricks or masonry or, in a few rare examples, it may be totally unlined. The water level in the well represents the water table (or phreatic surface).

Many boreholes are constructed with an upper section, which is lined with steel casing sealed into the ground with a concrete grout. This design serves two main purposes, that of providing support through unconsolidated or weathered materials near ground level, and also preventing surface water from entering the well and possibly contaminating the supply. Below the upper, cased section, the borehole is either unlined as shown in Fig. 6.1b or has a perforated well screen to allow groundwater to flow into the well. In both cases, the water level in the well represents the

(a)   (b)   (c)

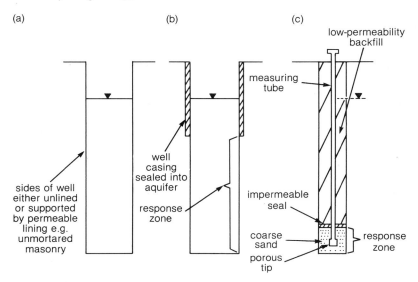

low-permeability
backfill

measuring
tube

well
casing
sealed into
aquifer

sides of well
either unlined
or supported
by permeable
lining e.g.
unmortared
masonry

response
zone

impermeable
seal

coarse
sand

porous
tip

response
zone

Fig. 6.1 Three different groundwater levels. (a) Uncased borehole/well. Water levels in the well can be taken to be the local *phreatic surface* (i.e. the water table). (b) Cased borehole. The water level represents the hydrostatic pressure (i.e. *pressure head*) in the uncased lower section of the borehole. This may be either a phreatic surface or a piezometric surface (i.e. the aquifer is confined), depending on the local hydrogeology. (c) Piezometer. The water level represents the hydrostatic pressure (i.e. pressure head) at a particular depth in the aquifer, determined by the level of the response zone. Except where the piezometer is just below the water table, this water level will not reflect the phreatic surface.

hydrostatic pressure in all the aquifer material penetrated by the lower section of the borehole. If the borehole has been drilled into a water table aquifer similar to that described in the case of Fig. 6.1a, then the water level in the borehole represents the phreatic surface. On the other hand, where the aquifer is confined by overlying low permeability deposits, the water level in the borehole will represent the piezometric surface when it stands above the top of the aquifer (Fig. 6.1b).

A third example (Fig. 6.1c), illustrates a further variation on this theme. A small diameter pipe has been installed in a borehole. At the bottom end of the pipe there is a perforated section or an unglazed ceramic pot which is surrounded with coarse sand. Above the sand there is an impermeable seal which is usually made from high grade clay such as bentonite (calcium montmorillonite), and the remainder of the borehole has been backfilled with a low permeability material. The water level in this pipe reflects the hydrostatic pressure in the aquifer at the level of the pipe bottom, and *not* the general phreatic surface in a water-table aquifer or the piezometric surface in a confined one. Such installations are usually referred to as *piezometers* and are often installed to examine groundwater conditions as part of a geotechnical investigation for civil-engineering construction. Sometimes several piezometers are installed at different levels in the same backfilled borehole. Such installations are generally referred to as a 'nest' of piezometers. These examples serve to illustrate the importance of knowing the construction details of a well or borehole where you are measuring groundwater levels. Without such information an interpretation of the levels is impossible.

6.1 Compiling a well catalogue

A well catalogue is a detailed list of all known wells and boreholes in an area, including both those in use and disused ones. In some instances it may be appropriate to include springs in this list. Even abandoned wells, which have been backfilled or covered over and lost, should be included. Each entry in the well catalogue should be identified with a unique number which will give easy access to the details for each site. The record should include identifying information such as the name and address of the site, the map reference, details of construction such as the drilling method, depth, diameter, casing and screen, information about the geology, water levels and abstraction rates. This information can easily be kept in a card index system, but nowadays microcomputer-based systems are becoming common and appropriate software is now commercially available.

Figure 6.2 shows a typical record card which could be adapted for most situations. All the relevant information needed for each entry can be summarized on cards 200×125 mm in size. It is fairly obvious from the titles what information should be written in each box on the card. Some additional notes are needed, however, to explain one or two headings.

6.1.1 Name of site

This should be the same name as is used in any other records (e.g. abstraction licence), but should include additional information if any difficulties are likely to arise in identifying the well (e.g. two wells on one site).

6.1.2 Abstraction licence number

In many countries, water abstraction is controlled by law, often by a licensing system. This space is for the abstraction licence number, so that cross-referencing is made easy. Often abstraction records and details of the well construction are kept by the abstraction licensing authority.

Name of site						Well catalogue No.		
						Geological survey No.		
Owner						Map grid ref.		
Occupier						Status		
Ground level		m OD			ft. OD	Aquifer		
Level of well top		m OD			ft. OD			
Rest water level		m bwt			ft. bwt	Summary of geological section	Thickness	Depth
(Date)		m OD			ft. OD			
Construction: Method			Date					

Depth bwt	Dia.	Linings (below well top)						
		From	To	Dia.	Type			

Abstraction rates (state units)		Type of pump					
	PWL	Chem./bact. anal.		YES/NO			
		Well driller					

If insufficient space has been allowed, continue in 'Notes' overleaf.

Name of site LOW GODDERTHWAITE FARM EAST WELTON						Well catalogue No. JF 80/404 A		
						Geological survey No. 27/128		
Owner MR J. HARRIS			Abstraction licence No. N/A			Map grid ref. JF 8097 0445		
Occupier E/WATER AUTHORITY						Status OBSERVATION BOREHOLE		
Ground level 47.98		m OD			ft. OD	Aquifer TRIASSIC SANDSTONE		
Level of well top 48.11		m OD			ft. OD			
Rest water level 40.83		m bwt			ft. bwt	Summary of geological section	Thickness	Depth
(Date 6/9/83)		m OD			ft. OD	ALLUVIUM	1.5	1.5
Construction: Method AIRFLUSH ROTARY Date SEPT. '82						BROWN CLAY	1.1	2.6

Depth bwt	Dia.	Linings (below well top)						
		From	To	Dia.	Type	RED SANDSTONE	12.4	15 M
0-5	100 MM	GL	5M	100 MM	PLASTIC	(NOT BOTTOMED)		
5-15	75MM					WATER STRUCK		
						AT 7.3 M		

Abstraction rates (state units)		Type of pump N/A					
—	PWL	— Chem./bact. anal.		YES/NO			
		Well driller BOREWOODS					

If insufficient space has been allowed, continue in 'Notes' overleaf.

Fig. 6.2 A record card, approximately 125×200 mm in size, can be used to summarize well-catalogue information. The cards can be filed in well-catalogue number order for quick reference. Additional information can be noted on the reverse side (see Fig. 6.3).

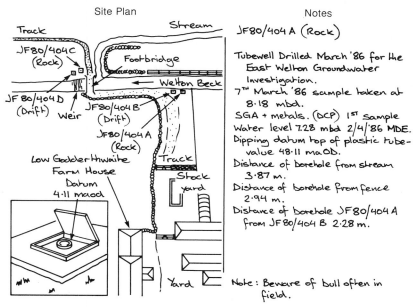

Site Plan

Notes

JF80/404 A (Rock)

Tubewell Drilled March '86 for the
East Welton Groundwater
Investigation.
7ᵀᴴ March '86 sample taken at
8.18 mbd.
SGA + metals. (DCP) 1ˢᵀ sample
Water level 7.28 mbd 2/4/'86 MDE.
Dipping datum top of plastic tube -
value 48.11 maOD.
Distance of borehole from stream
3.87 m.
Distance of borehole from fence
2.94 m.
Distance of borehole JF80/404 A
from JF80/404 B 2.28 m.

Note: Beware of bull often in
field.

Fig. 6.3 The reverse side of the record card can be used for a sketch plan to help locate the borehole in the future. Information such as the date when samples were taken and hazards such as bulls and dogs (or even a talkative farmer) can also be noted.

6.1.3 Geological survey number

In some countries a record of the geological information relating to each well and borehole is compiled by the Geological Survey. Recording the appropriate number again allows cross-referencing.

6.1.4 Well catalogue number

This is the identification number relating to your well catalogue. It may be a simple system which starts at '1'

and continues sequentially. This approach can be improved if part of the number is related to the map reference of the well or simply the topographic or geological map on which the well is located. For example, a number such as SJ27/129 may be split into map SJ27 and record number 129. Such a system allows more rapid access to data for card-based records

6.1.5 Map grid-reference

The topographic maps published for most countries have a grid-referencing system. If these do not apply, then latitude and longitude would be appropriate. It is suggested that the grid reference should be given as accurately as possible to identify the borehole to within 10 metres on the ground, if possible, and never more than 100 metres.

6.1.6 Status

This is the place to note whether the borehole is used for a water supply, is disused, or is a purpose-drilled observation borehole, etc. In comprehensive well catalogues which include springs and boreholes drilled for site investigation or mineral exploration, this is the place to record that detail.

6.1.7 Aquifer

Record the stratigraphic name of each aquifer here. The entry in this space enables details for all sources in a particular aquifer to be extracted rapidly from the record system. It is particularly useful where the well catalogue extends over a large area covering several different aquifers.

6.1.8 Notes and site plan

On the reverse of the card there is space provided for additional notes and for a sketch plan. It is often useful to draw a plan of the site, to help locate the borehole or well again. It is a good idea to also include a diagram of the head-works arrangements, showing the dimensions of important features and the reference datum used

for water level measurements. This helps when deciding which wells and boreholes are accessible and suitable for observation purposes and water sampling. Figure 6.3 shows a sketch of a site where boreholes could be easily confused, and provides enough information for each one to be identified.

The second part of the well catalogue should be a series of reference maps on which the position of every entry is marked, together with its identifying number. In Britain, a good scale map to use is 1:25 000, as the detail contained in Ordnance Survey maps of this scale is sufficient to enable locations to be identified with great accuracy. Where there is a high density of wells, it may be necessary to use a larger scale such as 1:10 000. In other countries you must select maps of an appropriate scale, from whatever is available.

When you are plotting the position of a well on to the reference map, you can read off the grid reference most accurately if you use a map roamer. An example is shown in Fig. 6.4, which is a type used by the British Geological Survey. It consists of a number of grids at different scales, printed on clear, hard plastic. Such a device is essential if accurate positioning is required.

6.2 Field surveys for wells, boreholes and springs

The first step in compiling your well catalogue is to obtain all available

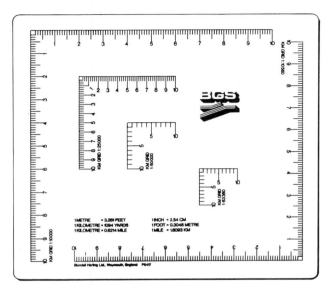

Fig. 6.4 A map roamer (BGS type).

information from existing records. Eventually, however, you will have to go out into the field to check details, measure water levels, and search for additional wells.

The best way to find wells in the field is to ask local people. The majority of villages in developed countries have a mains water supply, but outlying farms and houses will probably have their own source of water. The only way to find out is to knock on the door and ask. This requires diplomacy if the enquiry is to be fruitful. Be prepared to spend time explaining who you are and what you are doing. When undertaking this sort of survey, take a large-scale map along with you which shows field boundaries. In Britain 1:10 000 is the best scale. Ask each farmer or landowner to describe the boundaries of his property and then identify them on the map. Ask him to point out the position of all his wells, boreholes and springs. Do not expect everyone to be able to read a map, so visit each site with the farmer if necessary. He should also be able to tell you the name of his neighbours, so that your next visit will not be completely uninformed. In this way you will build up a picture of all the local sources, and and fields missed from your survey should be obvious.

You may have to be on the lookout for signs that a well or borehole exists. A hand pump is an obvious sign, but it

Field Hydrogeology

could be misleading as they are not always on wells. Hand pumps are also used to pump water out of underground rain water collection tanks, for example, so look under the manhole cover to check if it is a well or a tank. Wind-pumps are more easily seen, even when they are broken. They are generally sited over boreholes or wells and only rarely over rain water collection tanks.

Sometimes tall, 'thin' buildings may contain a borehole. The height is needed partly to house lifting gear to pull out submersible borehole pumps, and partly to house a water tank in the roof to provide sufficient head to drive water round a distribution system. Lifting tackle, in the form of a sturdy frame, may be permanently

installed over boreholes which are outside. Keep your eyes open. Eventually you will develop an instinct for finding wells and boreholes and will be able to spot them from your car as you drive by.

6.2.1 Locating springs

There are tell-tale signs to look for when locating springs. Your geologist's eye for the country will help to identify spring lines, such as the simple example in Fig. 6.5. Vegetation is a great help; rushes and sedges which grow in wet places are often a darker or lusher green than the grass covering the rest of the field or the hillside. Where springs are used for a water supply, be on the look out

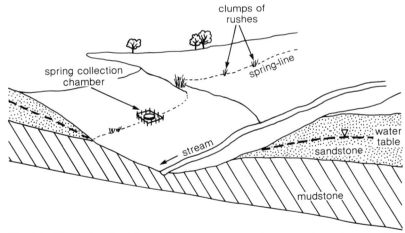

Fig. 6.5 The spring-line follows the contact between the permeable sandstone and the underlying impermeable mudstone. The surface signs are clumps of rushes (which are generally a darker green than the rest of the vegetation), the start of a minor tributary to the main stream and a spring-collection chamber. This latter looks like a concrete or masonry box, partially buried and fenced round (see Fig. 9.1).

72

(a)   (b)

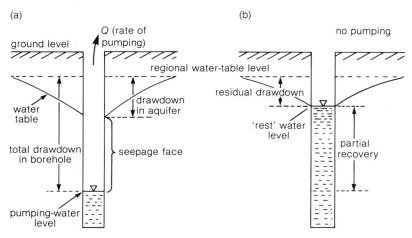

Fig. 6.8 Drawdown and recovery of water levels in an abstraction borehole. (a) The amount that the water level is drawn down below the regional water table is the sum of the drawdown in the aquifer and the height of the seepage face. The value of both components will vary with different pumping rates. (b) Once pumping is stopped, the seepage face quickly disappears, but recovery within the aquifer takes longer. As a result, water levels in the borehole are likely to be affected by residual drawdown for several days or even weeks after pumping has ended.

happens and, consequently, these devices must be regarded as suspect. It is best to avoid using them altogether and to use a dipper instead.

6.3 Interpretation of abstraction borehole water levels

If you are taking a water level in a supply borehole it is vital to note down whether it is a pumping or rest water level, and to take great care when interpreting this information. When water is pumped from a borehole, the water level is drawn down rapidly at first and then gradually

slows down until a stable pumping level is achieved. The amount of drawdown is largely controlled by four factors: the aquifer characteristics, the ease with which water flows through the well face, the rate of pumping and the length of time since pumping started. Fig. 6.8a shows the components of drawdown in an abstraction borehole. Note how the presence of a seepage face means that the water level in the borehole is drawn down to a much lower level than that in the surrounding aquifer. This difference is usually several metres or even several tens of metres. The water level in the aquifer is itself drawn down below the regional water level to form a *cone of depression*. When the pump is turned off, the water level in the borehole recovers very rapidly at first as the borehole fills up to a level equal to the aquifer water level. Recovery continues, but at a much slower rate as groundwater flows in to fill up the cone of depression (Fig. 6.8b). Full recovery back to the regional water level is likely to take several days, weeks or even months. These comments generally apply to water levels in boreholes which penetrate confined aquifers.

When comparing a record of pumping-water levels in a particular abstraction borehole, it is essential to know how long and at what rate pumping has taken place before each measurement was taken. If both pumping and rest water levels have been recorded over a long period, it is sometimes possible to estimate general trends in water-level fluctuations. Plot separate graphs of rest and

pumping levels as in the example shown in Fig. 6.9. This example shows a nine-year record of a factory borehole, where rest and pumping-water levels were recorded at monthly intervals. The rest water levels are fairly constant at some 30 metres below datum. This indicates that the pumping from the borehole is not causing a general decline in the regional water level. It does not provide a record of the regional level, however, as the rest levels only represent a partial recovery as shown in Fig. 6.8b. The few rest water levels which are higher than the others were taken after rest periods of a few weeks, instead of the usual two-day weekend shutdown. This suggests that the regional level is several metres higher than the rest water level record.

The pumping water level record varies, reflecting changes in the total quantities of water pumped from the borehole each year. The higher water

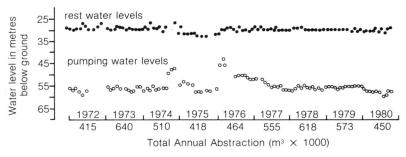

Fig. 6.9   The pumping-water record generally reflects changes in the total quantities pumped. The higher water levels in 1974 and 1976 were caused by a reduction in the hourly pumping rate. The rest water-level data show recovery to similar levels throughout the period of record, indicating that the abstraction has not caused a general lowering of groundwater levels in the area. The few higher levels occur after periods of resting which are longer than usual. (North West Water data.)

levels in 1974 and 1976 were caused by a reduction in the hourly pumping rate. In general, this record is useful information about both the behaviour of the abstraction borehole and the impact of the abstraction on the local groundwater levels. It is not adequate to provide much quantitative information, however, despite being more comprehensive than the majority of such records.

When using water-level information from abstraction boreholes, be very cautious. If long-term trends cannot be seen easily, it is better to reject the information than to be tempted into using your imagination too much. In some investigations there may be no real alternative to using this type of information. If you are forced to use abstraction boreholes as your main water-level observation points, try to standardize on the period you wait after the pump has been shut off, before taking a water-level reading. This period should be several hours at least, but it is likely that the demand for water and the need to recommence pumping will be the deciding factor in determining the length of this period. If you are using an industrial supply borehole then you will probably be able to increase the rest period in several ways, such as taking the reading just before the factory starts up again after a weekend shutdown, or during the annual shutdown during the holiday period. You should be able to manage a few hours rest from pumping by taking full advantage of the storage capacity in supply tanks.

6.4 Observation borehole networks

The different requirements for groundwater level measurements can give rise to conflicts in the design and operation of an observation borehole network. Such a network comprises the large number of measurement points which are needed to define groundwater level contours. In most instances, this means taking a measurement at all sites where it is possible, as well as noting the position and elevation of springs and those reaches of rivers and streams which are in contact with the aquifer. It is usually only necessary to take all these readings once or twice in a year, at the end of March or the beginning of April and in late October or early November when groundwater levels are at their maximum and minimum values respectively. (Note that the reverse will apply in the Southern Hemisphere.)

Monthly or even weekly readings on your network are needed to define the changes in groundwater levels in response to seasonal recharge and abstractions. The density of these observation points does not have to be as great as is needed for contouring water levels, and one borehole every 20 to 25 square kilometres of aquifer is usually adequate for regional studies. If you are examining a small area in detail, such as a river terrace, then you will need a greater density of observation points, perhaps even as great as one per hectare. The ideal frequency at which to take readings depends upon how quickly the aquifer responds to recharge, but it will normally take a year or two to find this out. As a rule of thumb, for general monitoring purposes, take monthly readings and install water-level recorders on no less than ten per cent of the boreholes. If you are carrying out a monitoring programme associated with a pumping test or to examine the effects from a particular well or group of wells on a ground-water regime, then more frequent readings will be needed. Each case will be different, but often two or three readings each week are ideal. Practical considerations usually make monitoring at these frequencies very time consuming. A workable compromise may be to fit part of your measurement programme in with other work in the same area. Keep a dipper in your vehicle and measure water levels when you are passing any of your observation boreholes, thereby obtaining valuable extra data with minimum effort. When the readings are plotted up as a hydrograph, it will be possible to interpolate values between each reading to avoid any problems caused by the variation in the frequency between measurements, or differences in the times when readings were taken at different sites.

6.5 Groundwater level fluctuations

In most groundwater studies you will be examining groundwater level data to determine seasonal fluctuations in response to recharge, or to see if pumping is causing levels to decline and water supplies to fall. It is important to understand how other factors can cause water levels to change. The theory behind these changes is described in all standard groundwater textbooks, such as those listed in the bibliography. Broadly, these changes can be divided into four groups, as outlined below.

6.5.1 Changes in groundwater storage

These are caused by such factors as increases or decreases in abstraction and seasonal rainfall. Figure 6.10 is a well hydrograph which illustrates these changes.

The water level in aquifers which are in hydraulic contact with a river, will fluctuate in response to changes in river level as water flows in and out of the river-bank deposits. The extent of these changes will depend upon the size of the river-level fluctuations and the permeability of the bankside and aquifer materials. Comparison of well hydrographs with the river hydrograph (see section 7.7) will help to detect this relationship.

In coastal aquifers which are in direct hydraulic contact with the ocean, a similar effect is observed. The semi-diurnal nature of these water-level changes and the relationship of these with the tide cycle will help you decide if an observation borehole is responding in this way.

Other possible factors which may affect the quantity of groundwater in

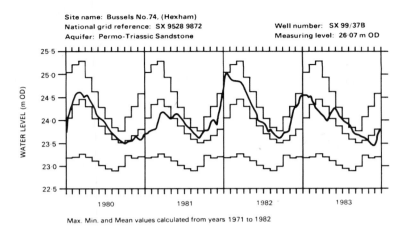

Site name: Bussels No.74, (Hexham)
National grid reference: SX 9528 9872
Aquifer: Permo-Triassic Sandstone

Well number: SX 99/37B
Measuring level: 26·07 m OD

Max. Min. and Mean values calculated from years 1971 to 1982

Fig. 6.10 The graph shows groundwater levels recorded on an observation borehole in Devon, England. The dark line is the record for the period as shown, while the stepped graphs are the mean, maximum and minimum monthly values for the water level record to-date. The effect of seasonal recharge from September to February is obvious, but note how there are marked variations from year to year. (Reproduced from Hydrological Data United Kingdom 1984 Yearbook by permission of the Institute of Hydrology.)

storage, include leaks from water-supply pipes and sewers, both of which are common in urban areas throughout the world. The development of new urban areas and an associated increase in paved areas is likely to reduce recharge and cause a lowering of water levels, especially where surface drainage is diverted to sewers. Evapotranspiration from deep-rooted plants or in areas with a shallow water table will lower the water levels, particularly during the summer. Crop irrigation, on the other hand, is likely to artificially increase recharge and will cause water tables to rise even to the point of waterlogging soils and surface flooding. With the exception of responses to tides and river levels, all these changes happen slowly over a period of years.

6.5.2 Barometric pressure

Changes in atmosphere pressure produce significant fluctuations in wells in confined aquifers. As the water level in such a well is in contact with the atmosphere, it represents a balance between the hydrostatic pressure in the aquifer and the barometric pressure. In consequence, changes in atmospheric pressure produce a corresponding change in water level, with the water level falling as barometric pressure rises and rising as barometric pressure drops. A large

81

number of observation boreholes exhibit this behaviour, even some which penetrate water table aquifers. Generally, in water table aquifers, changes in atmospheric pressure are transmitted equally to the water table in both the aquifer and in a well. Hence no pressure differences occur. Air trapped in pores below the water table, however, responds to these pressure changes and causes similar but smaller fluctuations in water level. Permeability often varies vertically in an aquifer under the influence of sedimentary controls (see Chapter 4) and this may produce a confining effect, particularly in deep aquifers which permit the effects of barometric pressure changes to be reflected in observation borehole water levels.

When you are interpreting a groundwater-level record which is affected by barometric changes, it should be corrected to enable other influences to be clearly seen. To do this, it is first necessary to calculate the *barometric efficiency* of the aquifer. This must be done for each borehole as local geological conditions can cause it to vary from place to place. Barometric efficiency is simply the change in water pressure, divided by the change in barometric pressure expressed as either a percentage or a decimal. The two records must be expressed in the same units. This can easily be done by converting the barometric readings to centimetres of water. Figure 6.11 shows one method of calculating barometric efficiency, involving plotting the two records so that they can be compared visually. It

is important to remember to plot the increasing atmospheric pressure in a *downward* direction. In this example the change in water level and its corresponding change in atmospheric pressure was estimated from the graph (see caption to Fig. 6.11). An alternative method is to plot a graph of barometric pressure on the x-axis and water levels on the y-axis. Figure 6.12 was prepared using the data from Fig. 6.11. The generally poor fit of this plot is attributed to the borehole being situated some 18 kilometres from the meteorological station where the barometric measurements were taken. This will cause a time lag between the two records, which will vary depending upon prevailing weather conditions.

Correction of groundwater level data for barometric changes is most important during pumping tests conducted to determine aquifer properties. In these circumstances the barometer should be situated close to the test borehole site. Experience has shown that the majority of barometric efficiency values fall in the range of twenty to eighty per cent. To correct your groundwater level for barometric efficiency effects, follow the steps given below:

1 Determine the atmospheric pressure at the time each groundwater level was measured.
2 Convert these values to units compatible with your level record, (e.g. cm of water, where 1 000 m_{bar} = 1019.7 cm of water at 4°C).

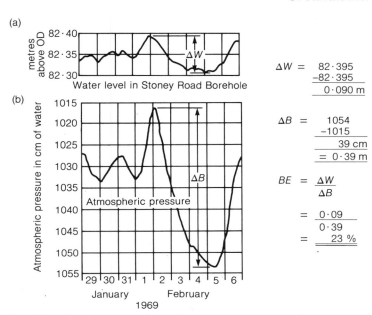

Fig. 6.11 Groundwater levels (a) fluctuate in response to changes in atmospheric pressure (b). Note that the vertical scale of each graph is the same but increase in opposite directions.

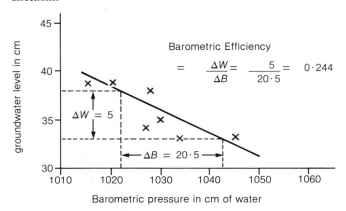

Fig. 6.12 Barometric efficiency can be calculated by plotting water levels against atmospheric pressure (expressed as a column of water). The slope of the straight line is the barometric efficiency expressed as a decimal or as a percentage. (Note: based on data used in Fig. 6.11.)

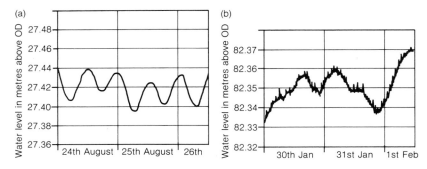

Fig. 6.13 The three graphs are copies of recorder charts on observation boreholes and show various features of such records. (a) This borehole is in a confined aquifer and lies within 1.5 km of the coast. The regular fluctuations which occur at 12-hourly intervals are caused by the tide. The changing load of sea-water on the aquifer causes pressure changes within the aquifer which, in turn cause a change in level. In this example, high tide occurred approximately three hours before the peak in water level. (b) This borehole is in a confined aquifer and is located at a railway station. The very noisy trace is caused by changes in loading on the aquifer as trains pass by. The overall changes in water level are caused by fluctuations in barometric pressure, data from this borehole also being shown in Fig. 6.11. (c) This borehole is in a confined aquifer and lies within 2.5 km of the coast in northwest England. The record of water levels shows the effect of tidal movement and barometric fluctuations. The main feature on this trace, however, is a fluctuation in water level of some 12 cm, caused by shock waves from the Mexican earthquake on 19 September 1985. (North West Water data.)

3 Calculate the difference in atmospheric pressure between each reading and the previous one. The values obtained should be multiplied by the barometric efficiency. This will give the size of the water-level movement caused by the barometric change.

4 If the atmospheric pressure has increased between the time of one water level reading and the next, *subtract* the amount calculated in 3 from the measured drawdown, where the water-level reading is measured as a *depth* below the ground-level. This is because the increasing atmospheric pressure will have depressed the groundwater level. The reverse applies when the barometric pressure has fallen. Here, the correction value should be *added* to field readings of drawdown. Where groundwater levels are corrected to read *above* a datum (e.g. sea-level), reverse the procedure adding the correction when the atmospheric pressure increases and *vice versa*.

(c)

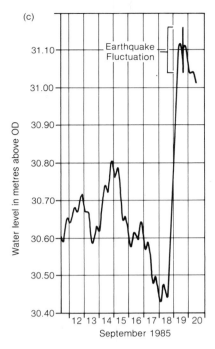

September 1985

Figure 6.13a shows a recorder chart from a tidally affected borehole. A similar effect is caused by large point loading of the aquifer, the most common example of this being provided by the passage of railway trains. As the train approaches an observation borehole in a confined aquifer, the well water level rises in response to the increased load. Once the train has passed, the water level falls below the undisturbed level, before returning to rest again. These changes occur over a very short period, the same time in fact, that the train takes to pass by. As a consequence, the passage of trains usually only shows up as 'noise' on recorder-chart traces (see Fig. 6.13b). The amount of change in a groundwater level is rarely more than a few millimetres, and so is rarely of more than a curiosity once the cause of the noisy trace has been identified.

6.5.3 Changes caused by loading

Confined aquifers have an elastic property which results in changes in the hydrostatic pressure when changes in loading occur. In coastal areas, tidal fluctuations cause a large variation in the mass of sea-water on the aquifer, which causes a corresponding change in pressure within the aquifer. These tidal fluctuations are direct, that is to say, as sea-level increases so does the hydrostatic pressure and the water levels in the wells rise. These effects can be recognized by their semi-diurnal nature and by comparison with tide tables.

6.5.4 Miscellaneous

There are a variety of other causes for groundwater level fluctuations. Earth tides are movements of the Earth's crust in response to the gravitational pull of the Moon and, to a lesser extent, the Sun. These movements have been shown to produce semi-diurnal water level fluctuations in confined aquifers, and can be at any distance from the coast.

Earthquakes can cause changes in groundwater levels in a variety of ways. In the area directly affected by the earthquake, spectacular changes can occur, such as the appearance and disappearance of springs, fluctuations

Field Hydrogeology

in spring discharges and rises or falls in groundwater levels. A permanent rise in the water table can be caused by Earth tremors inducing further compaction of non-indurated sediments, thereby reducing the storage capacity of the aquifers. Most frequently, however, earthquake shocks cause small fluctuations in the water level in wells penetrating confined aquifers. Such responses can be seen in boreholes on the other side of the world from the earthquake epicentre (see Fig. 6.13c). A similar effect on water levels is caused by the underground testing of nuclear weapons.

The trace of a recorder-chart can fluctuate for many reasons other than changes in water level. Wind gusting over the open top of a borehole causes a sudden drop in air pressure inside the casing, which then produces an immediate rise in the water level. Strong winds can give rise to a 'noisy' chart more directly by rocking poorly secured recorder cabinets. Animals can also affect the water level record. For example, cattle or horses may use a recorder cabinet as a scratching post, thereby rocking both cabinet and recorder, causing an unexplained wiggle of the pen trace. If this is done by cows on the way to being milked, a regular semi-diurnal feature will result on the record, causing great confusion to the hydrogeologist. Should any small animal, such as a frog, fall into a well which contains a recorder, it is likely to climb on to the float. All movements of the animal will be transmitted to the recorder, again causing confusion.

6.6 Constructing groundwater contour maps and flow nets

Before you start to interpret groundwater-level data it is very important to decide whether all your measurements relate to the same aquifer. Examine borehole construction details and the geological structure. Where adequate data are not available, compare water level fluctuations. Generally, fluctuations follow the same pattern throughout an aquifer.

Groundwater level data can be used to determine the direction of groundwater flow by constructing groundwater contour maps and flow nets. A minimum of three observation points are needed to calculate a flow direction. The procedure is first to relate the groundwater field levels to a common datum – map datum is usually best – and then accurately plot their position on a scale plan, as in Fig. 6.14. Next, draw a pencil line between each of the observation points, and divide each line into a number of short, equal lengths in proportion to the difference in elevation at each end of the line. In the example shown in Fig. 6.14, each division in the lines A–B and B–C is 0.2 metre, while on line A–C each division is 0.1 metre. The length of each division on the plan is not related to the scale of the plan.

The next step is to join points of equal height on each of the lines to form contour lines. Select a contour interval which is appropriate to the

86

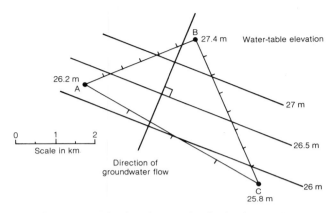

Note: The water-table elevation values must be related to the same datum.

Fig. 6.14 Estimation of groundwater contours and flow directions, based on water levels in three observation boreholes.

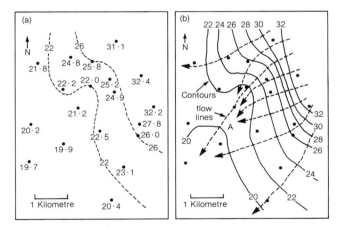

Fig. 6.15 Construction of a groundwater contour map and flow net. (a) Values of groundwater levels are located on a plan, and key contours are plotted using the techniques shown in Fig. 6.14. In this example contours at 22 m and 26 m were used. (b) The remaining contours are interpolated in the same way. Flow-lines were sketched in, perpendicular to the contour lines, starting on the 30-metre contour at a spacing of 500 m.

overall variation in water levels in the study area. The direction of ground-water flow is at right angles to the contour lines.

This simple procedure can be applied to a much larger number of water level values to construct a groundwater-level contour map such as the one in Fig. 6.15. First, locate the position of each observation point on a base-map of suitable scale, and write the water level against each position. Study these water level values to decide which contour lines would cross the centre of the map. Select one or two key contours to draw in first.

In Fig. 6.15a the 22-metre contour was drawn by interpolating its position between each pair of field values, following the procedure illustrated by Fig. 6.14. Once this was completed, the 26-metre contour was drawn in the same way. The remaining contours were drawn using the two key contours as a guide, in addition to interpolation between the field values.

Once the contour map is complete, flow lines can be drawn on by first dividing a selected contour line into equal lengths. Flow lines are drawn at right angles from this contour, at each point marked on it. The flow lines are extended until the next contour line is intercepted, and are then continued at right angles to this new contour line. In the example shown in Fig. 6.15b, the 30-metre contour was divided into 500 metre intervals as the starting point for constructing the flow lines. Always select a contour which will enable you to draw the flow lines in a downslope direction.

It is important to bear in mind what you have already learned about a particular aquifer while you are constructing a groundwater-level contour map and flow net. Information on the geological structure and variations in aquifer hydraulic properties should be taken into account. This is particularly important where groundwater flow is through fissures, karstic limestones being an extreme example. It is not unusual in such aquifers to find that the fissure systems do not permit flow in the directions indicated by the contours and real flow directions can only be deduced using tracers (see section 7.9).

A groundwater level contour and flow line map represents groundwater movement in a plan form only, and therefore is only part of the picture. When viewed in cross-section, groundwater flow-paths curve towards a discharge point such as a spring-line, stream or even a pumping well. Figure 6.16 shows two examples of groundwater flow at depth in an aquifer. It is important to try and envisage the three dimensional flow paths which groundwater is following in the aquifers in your study area.

6.7 Interpretation of contour maps and flow nets

A groundwater contour map can represent either a water table (phreatic surface) or a piezometric surface, and this should be sorted out from geological and well construction in-

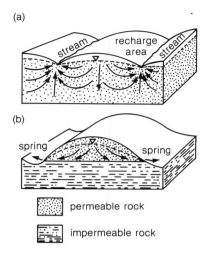

(a)

stream recharge area stream

(b)

spring spring

[permeable rock] permeable rock

[impermeable rock] impermeable rock

Fig. 6.16 The direction of groundwater flow at depth is not usually parallel to the water table; instead, water moves in a curved path, converging towards a point of discharge. In (a), the ground is uniformly permeable and groundwater discharges into streams along the valleys; it may approach the stream from the sides or from below. In recharge areas, this downward flow would be reflected in the relative levels in a nest of piezometers, where the water level would be highest in the upper piezometer and lowest in the deepest one. In discharge zones the reverse is true, and this accounts for flowing boreholes being common in river valleys. In (b), the hill is capped by a permeable rock which is underlain by an impermeable stratum. Groundwater is diverted laterally by the impermeable material, and springs result at the ground surface–along the contact between the permeable and impermeable strata. (Reproduced from S238 by permission of the Open University.)

formation. It is possible that one part of an aquifer is confined while the remainder has a water table. Indeed, such complexities are common but must be understood before ground-water contour maps and flow nets can be interpreted.

The spacing of groundwater con-tours gives a good indication of vari-ation in aquifer permeability values. Where contours are close together it is indicative of low permeabilities, be-cause a steep hydraulic gradient is needed to 'push' the water through the aquifer. Where groundwater con-tours are more widely spaced, the converse is true, and the aquifer is likely to be much more permeable.

Groundwater flow-lines indicate not only the overall direction of flow but where flow is concentrating. This is also an indicator of variations in aquifer permeability. In the example in Fig. 6.15b, the groundwater con-tour spacing suggests that the south-west of the area of the aquifer is more permeable than the northeast. The flow lines show that there is a concen-tration of flow in the area marked 'A', and this is probably an area of discharge to surface streams. If this map were to be used to select favour-able locations for a new well, area 'A' would have much to commend it as it is both in an area of high permeability values and where flow lines converge.

Water-level change maps (Fig. 6.17) are used to calculate changes in the volume of water stored in an aquifer as part of a water-balance exercise (see Chapter 8). These maps are also useful when

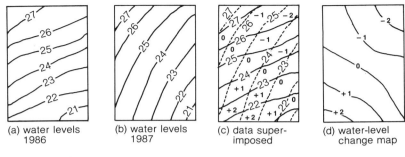

(a) water levels
   1986

(b) water levels
   1987

(c) data super-
   imposed

(d) water-level
   change map

Fig. 6.17 Changes in groundwater levels can be easily studied if you construct a water-level change map. In this example, groundwater-level contour maps (a and b), covering the same area in successive years, are superimposed. The differences in values are noted at points where the contours cross (c), and new contours are drawn to show the amount of change (d). Water-level change maps are useful to measure the local effects of recharge and discharge. In this example the recharge area lies in the southwest corner and the discharge area is in the northeast, facts which are not obvious from the groundwater-level contours but become apparent once the water-level change map is constructed.

assessing the local effects of recharge and abstraction.

6.8   Using other groundwater information

In many groundwater studies there are fewer groundwater level measurement points than the hydrogeologist would like. The construction of new boreholes to provide this extra information is very costly, and so most experienced hydrogeologists have learned to be expert in the use of indirect information on groundwater levels. Observation borehole records can be supplemented by using information on the position and elevation of springs, *provided* that there is field evidence to show that these springs

discharge from the same aquifer that the boreholes penetrate. This evidence could be geological – such as the spring issuing from the same formation. A study of groundwater chemistry information can often help determine whether or not a spring issues from a particular aquifer. This is discussed in more detail in section 7.8.

In addition to springs, streams and rivers can also be used to help construct a groundwater contour map. Field evidence is again needed, and this may include outcrops of the aquifer in the stream bed, and a significant increase in flow as the stream flows across the aquifer. Other information commonly used includes topographic details, usually taken from published maps. If the study area

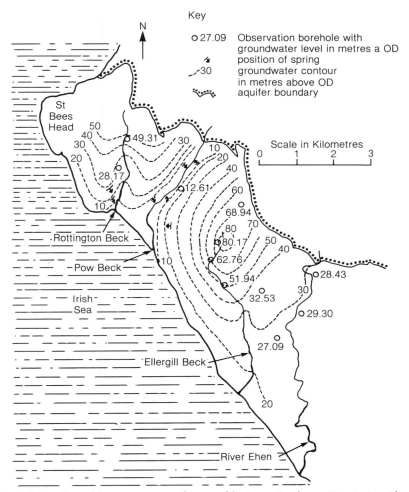

Key

○ 27.09   Observation borehole with
             groundwater level in metres a OD
             position of spring
  -30   groundwater contour
             in metres above OD
             aquifer boundary

Scale in Kilometres

0   1   2   3

Fig. 6.18 Groundwater contour map for part of the St Bees Sandstone (Permian) aquifer in West Cumbria, UK (see Fig. 4.2). The aquifer lies unconformably on older rocks (mainly mudstones) and dips to the west. Areas of high land are generally at outcrop and form the main recharge areas. The contours were drawn using water-level data from 11 observation boreholes–the elevation of springs which are known to discharge from the main aquifer, where streams (becks) and rivers flow over the outcropping sandstone and the general topography of the area. This latter information was obtained from published contoured survey maps for the area. Note the contrast in the steepness of the groundwater 'surface' and the variation in the distribution of data points. (North West Water data.)

is on the coast, it is usual to assume that the coast approximates to the zero groundwater contour. Similar assumptions can be made in respect of any large lakes in more inland locations.

Figure 6.18 shows the groundwater contour map, drawn up as part of a groundwater resource appraisal of the St Bees Sandstone aquifer in West Cumbria, England. Groundwater levels were only available at 11 sites which were not evenly distributed over the 30 square kilometres of aquifer. This information was supplemented by a number of springs which all lie in the northern part of the aquifer. Similarity between the chemistry of the water from these springs with that of groundwater samples taken from the observation boreholes has demonstrated that the springs drain from the main aquifer and not the overlying drift material. Flow

gauging (see section 7.4) was carried out on the four streams which cross the aquifer, to ascertain where groundwater discharge points lie. This information was taken into account when the contour map was constructed, together with the shape of the local topography and an assumption that the coast approximates to the zero contour. Despite this additional information, it was not possible to draw contours with a more frequent interval than ten metres, and there was insufficient detail to allow the 10-metre contour to be drawn over much of the aquifer area. Nevertheless, even with these limitations, the contour map indicated that the northern part of the aquifer would be unlikely to be suitable for production boreholes, in contrast to the southern end. These conclusions were supported by subsequent pumping tests.

7

Rainfall, springs and streams

This chapter describes how to measure the components of the hydrological cycle other than groundwater. Earlier chapters will help you to plan the fieldwork programme and decide which parameters should be measured. In addition to taking your own measurements, you may be using those taken by others. Make it a rule to inspect rain gauges or river-flow gauging stations which provide these data. It is important to assess how accurately the readings are taken. In addition, walking the river-bank may reveal otherwise unsuspected abstractions or diversions which could have a significant effect on your field readings.

7.1 Precipitation

In most parts of the world, the majority of precipitation falls as rain which is measured by a network of rain gauges operated by national and local government agencies. It may be that one of these rain gauges is sited within your study area or close enough for you to use these records. This is not always the case, and it may be necessary to set up a rain gauge for the duration of the study. Sometimes,

even when a rain gauge is available close to the study area, extra gauges may be needed to examine local variations which can be significant in areas of high relief, for example.

If you cannot get hold of a standard rain gauge (Fig. 7.1) you can make an improvised one using a flat-bottomed glass bottle and a funnel. A possible arrangement is shown in Fig. 7.2. An improvised instrument is not likely to be as accurate as a standard one, but it will be adequate for most hydrogeological purposes.

The amount of rain collected in a rain gauge is measured using a calibrated cylinder. This is a narrow cylinder which is graduated in millimetres of rainfall and is related to the rain gauge funnel diameter. This system enables readings to be made which are accurate to one-tenth of a millimetre. It is standard practice to record the daily total as a depth in millimetres to the nearest 0.1 mm *at or below* the bottom of the meniscus. If the volume of rainfall is less than 0.1 mm it is recorded as a trace. Records are kept as daily totals, with measurements being taken at the same time each day. In Britain this is traditionally at 9 a.m. GMT, with the

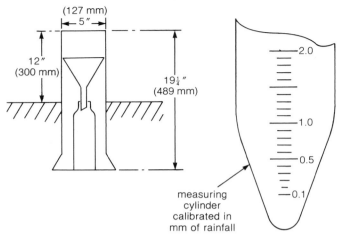

Fig. 7.1 This example of a 'standard' daily rain-gauge is a type used by the British Meteorological Office. It consists of a brass cylinder which incorporates a funnel leading into a collection bottle. The instrument is set into the ground to prevent it being knocked over and to keep the collected rain cool, thereby reducing evaporation losses in warm weather. Measurements are taken at the same time each morning, and the daily total attributed to the *previous* day and recorded as shown in Fig. 7.3. A special measuring cylinder is used to measure the rainfall to 0.01 mm.

total being recorded as the rainfall for the *previous* day.

Once you have measured the volume of rain collected in the gauge, shake out any remaining drops of water from the bottle. It is not usually necessary to dry it. If it is raining when you take the measurement, empty the collecting bottle into the measuring cylinder as quickly as possible, and then return it to the rain-gauge. Any rain that you miss during that period is unlikely to be significant. Remember that you are recording the volume of rain which falls during a fixed period of 24 hours, so that the remainder of that shower will be logged with the rain which falls the next day.

If you miss reading the rain-gauge for a day or two, try to maintain as accurate a record as possible. Record 'nil' rainfall if you know or it is quite obvious that there has been no rain (or snow, hail, etc.). Enter 'tr' for trace, if you know that there has been some rain, even if the bottle is bone-dry when you take the reading. Record any collected rain as an accumulated total for the days since you took the last reading. Bracket the days together on your record sheet, but make a note of any showers. See Fig. 7.3.

It is not worth trying to make a measuring cylinder which is calibrated in millimetres of rainfall for a home-made rain gauge. It is better to measure the volume of collected rain-

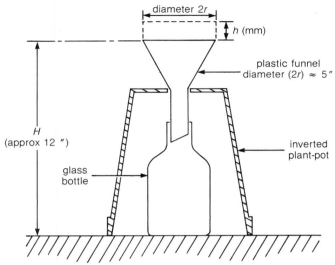

Fig. 7.2 This improvised rain-gauge uses a plastic funnel, glass drinks bottle and a large plant-pot. The volume of rain-water collected in the bottle is measured using a measuring cylinder, to the nearest millilitre. Rainfall is expressed as a depth in millimetres and is calculated from the formula:

$$h = v/10\pi r^2$$

where h = rainfall in *millimetres*, v = volume of rain-water in *millilitres* (which are the same as cm^3), r = radius of the funnel in *centimetres*. Record the rainfall in mm, rounded *down* to the nearest 0.1 mm.

Stn. name East Side Farm   Month March   Year 1987

Enter amount measured at 9h GMT against YESTERDAY'S date

Date	mm	Enter time of measurement if not close to 9h GMT and notes on significant weather	Date	mm	Enter time of measurement if not close to 9h GMT and notes on significant weather
1	2.0		16	1.1	
2	TRACE		17	5.7	HAIL
3	1.7	SNOW	18	0.6	
4	TRACE		19	1.0	
5	3.3		20	2.4	SNOW
6	7.1	RAIN / SNOW	21	—	
7	7.0	SNOW	22	5.4	SLEET IN AM.
8	—	THAW MIDDAY	23	4.3	SNOW BEFORE DAWN
9	—		24	10.6	HEAVY SHOWER LATE AM.
10	—		25		
11	—		26		SLEET IN AFTERNOON
12	TRACE		27	8.0	(Acc. Total).
13	TRACE		28	0.1	
14	0.5		29	0.6	
15	0.3		30	—	
			31	4.3	
OBSERVER H. Carson			TOTAL	66.0	

Fig. 7.3 In this example of how to record rainfall field data, notes have been made of significant precipitation events. No readings were taken for the 25th and 26th; instead the rainfall for these days has been included with that for 27th as an accumulated total.

water in a measuring cylinder, and then to convert this volume into a depth using the method shown in Fig. 7.2. When calculating the conversion factor do not forget that *millilitres* are equivalent to *cubic centimetres*, and that you want to record the rainfall as a depth in *millimetres. If you get it wrong you will end up being out by a factor of 10.* The volume of rain-water should be measured to the nearest millimetre, which means that a laboratory-type measuring cylinder should be used. It is possible to buy small plastic ones from suppliers of gardening equipment or shops which sell wine-making equipment.

Siting a rain-gauge is very important. An ideal site is a compromise between being out in the open, so that the gauge is not sheltered by trees or buildings, and exposed sites where strong winds are likely to cause near-ground turbulence which could blow raindrops out of the mouth of the gauge. A rain-gauge should be set at a distance from trees and buildings which are equivalent to at least twice their height, i.e. the vertical angle between the gauge and the top of the trees or building should be no more than 30° (Fig. 7.4).

Snowfall is recorded as an equivalent depth of rain, i.e. in mm of water. It is measured by collecting a representative depth of snow in a cylinder and then carefully melting the snow and measuring the volume of water produced. Do not heat the snow on a stove so that you lose water by evaporation. Stand the collecting

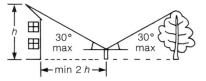

Fig. 7.4 A rain-gauge must be sited so that measurements are not affected by turbulence caused by the proximity of buildings or trees.

vessel in a bowl of warm water and allow it to melt slowly. When taking a sample, be careful to avoid both small accumulations and 'thin patches' if the snow has been drifting, and do not collect snow left lying around from earlier falls. These problems indicate that snowfall measurements are difficult and subject to more errors than rainfall measurements. If you are studying an area where snow forms a significant part of the total precipitation, a snow survey will have to be carried out. This is a systematic measurement of snow thickness and densities (i.e. rainfall equivalents) carried out along traverse lines. The information so gathered can be used to calculate the total volume of water represented by the lying snow.

7.2 Evaporation

Evaporation measurements rarely form part of a hydrogeological field investigation in countries where published figures are available, as these are usually adequate for water balance purposes. Where these data are not easily available, it will be necessary to make your own measurements. It is not possible to measure

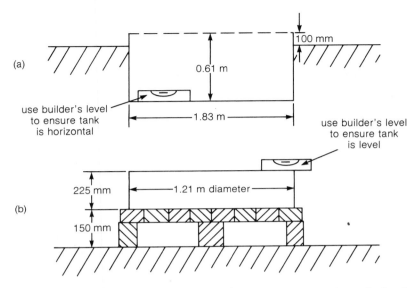

Fig. 7.5 Two types of evaporation tank in common use are the British Standard tank and the US Class-A evaporation pan. (a) The British Standard tank is 6 feet (1.83 m) square and 2 feet (0.61 m) deep, is made of galvanized iron, and is set into the ground with the rim 4 inches (100 mm) above ground level. (b) The US Class-A pan is circular with a diameter of 47.5 inches (1.21 m) and is 10 inches (225 mm) deep. It is set 6 inches (150 mm) above the ground on a wooden platform so that air can circulate round it. In both cases the tanks are filled with water, which is kept topped up. The water level in the US version is kept about 50 mm below the rim, while the water level in the British tank is not allowed to fall lower than 100 mm below the rim. Both types suffer from excessive losses, especially in hot climates where the water is likely to overheat. See Fig. 7.6 for an example of how to record these measurements.

evaporation except as losses from open water. As these losses are the maximum which can occur, they will only give an idea of the potential losses from other surfaces such as soil or from plants.

Two types of evaporation tank are in general use, although a number of others are also to be found. The two common ones are the British Standard evaporation tank and the US Class A evaporation pan (Fig. 7.5). Evaporation losses are calculated by carefully measuring the water level in the tank at the same time each day, and allowing for any water added to the tank to keep it topped up and also for any rain which may have fallen into it. Evaporation losses are expressed as rainfall equivalents, i.e. a depth in mm. Figure 7.6 provides an example of a daily record of evaporation losses.

August	Observations made at 0900 GMT		Daily records	
	Rainfall (mm)	Evaporation tank water level (mm)	Rainfall (mm)	Evaporation (mm)
1	1.2	23.6	—	1.5
2	—	22.1	10.2	0.7
3	10.2	31.6	—	1.8
4	—	29.8	8.1	2.3
5	8.1	35.6	—	1.6
6	—	34.0	0.4	2.0
7	0.4	32.4	*	*

* Values not included

Fig. 7.6 Daily readings of rainfall and evaporation-tank water levels have been taken at 09.00 GMT. Rainfall is attributed to the previous day and the difference in tank-water levels is used to calculate the evaporation losses. For example, there was no rain on 1 August so evaporation is calculated as follows:

tank level on 1st 23.6
tank level on 2nd 22.1
∴ fall due to evaporation = **1.5 mm**

When rain has fallen this must be taken into account as follows:

tank level on 5th 35.6
tank level on 4th 29.8
∴ rise in level = **5.8 mm**
rainfall on 4th 8.1
rise in level 5.8
∴ evaporation = **2.3 mm**

If you need to take your own evaporation measurements and you are unable to obtain one of the standard tanks, it is not too difficult to make your own. It needs to be about the size of the standard tanks described above, to help reduce problems of the water heating up and thereby increasing losses. This happens more easily with small volumes of water, and means that old baths and oil drums should be avoided.

Perhaps you might be able to find a suitable tank such as one intended for an ornamental pond or one sold by an agricultural supplier for animal watering. It is important to have one with vertical sides, so that the water surface area remains constant and the calculation of water losses is kept simple. The tank must be set with the rim absolutely horizontal, so use a builder's level, because the measurements use the rim as a datum point.

Make sure that birds and animals do not use it as a watering hole! This is likely to be a problem even in temperate climates, and you will need wire netting to cover the tank and a fence to keep the animals away.

7.3 Springs

The simplest way of measuring the flow of most springs and even some small streams is the 'jug and stop-watch' method. To do this, place a small vessel of known capacity below the spring, and time how long it takes for it to fill. It is essential to have all of the flow going into the vessel and to make sure that none of the water splashes out. With care, this method can be very accurate.

As the name implies, a very popular calibrated vessel for this job is a kitchen measuring-jug. It is best to get a jug with a capacity of one litre and made of polythene. If the jug is made of brittle plastic it will soon crack and spoil the measurements. Try to select a short, fat jug rather than a tall, thin one, as there may not be much space for the jug to fit under the falling water. Accurate measurements need the jug to be upright, so tall ones may not be suitable. It is a good idea to make the measuring mark easier to read by painting over it or sticking on coloured plastic tape. Electrician's insulation tape is ideal for this purpose.

When the spring is running so fast that if fills a jug in less than five seconds, a bigger container such as a 10 litre (two gallon) bucket should be used. Again, try to choose one that is short and fat and made of robust material. The black polythene buckets sold by most builder's merchants are very good but probably will need to be calibrated. This is quite simple to do, using the measuring-jug and marking the 10-litre level clearly with paint or tape. Put two marks on opposite sides of the bucket– it will make it easier to tell if you have kept it level during the measurement.

It may be necessary to modify the spring before you can go ahead and use your jug and stop-watch. If you intend to use this method to measure the flow of a small stream, you will certainly need to carry out some modifications. It is easiest to use this method when all the flow has been diverted through a short length of pipe. There must be sufficient gap below the end of the pipe to insert the jug or bucket and stand it upright. This can be achieved by building a small dam of stones and clay or concrete, through which the pipe projects for at least 200 mm. The important thing is to make sure that there are no leaks and that all the water goes through the pipe. You may have to dig a shallow hole beneath the pipe to make sure that the bucket will fit. Place a flat stone or a concrete slab beneath the pipe for the water to fall on to, to prevent bed erosion (see Fig. 7.7).

If the spring is already piped into a cattle drinking-trough or a catchpit which forms part of a water-supply system, measure the flow as it falls

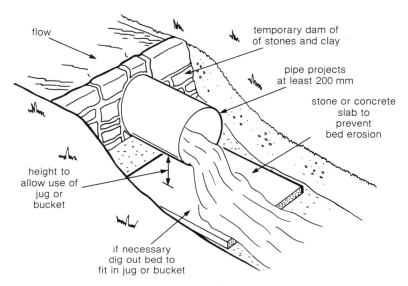

Fig. 7.7 Modify springs so that their flow can be measured with a jug and stop-watch by installing a temporary dam and diverting the flow through a pipe. Ensure that the pipe is high enough and long enough to allow the jug or bucket to fit underneath.

from the overflow pipe. If this is not possible, bail the tank out first so that you can get the jug or bucket below the inlet pipe. When taking measurements on a drinking water supply system, ensure that your hands and the equipment are perfectly clean, and take care not to knock any soil or other debris into the catch-pit (see section 10.7).

When taking a reading, bear in mind that it is important for the jug or bucket to be upright, otherwise it will be difficult to decide when the water level has reached the mark. Ensure that the stop-watch is started at the same instant that the first drop of water falls into the bucket, and stop it as soon as the mark is reached. Repeat

the measurement at least three times, and take the average as the correct flow. If there is a large discrepancy between the three readings, taking a few more readings will help to reduce errors (Fig. 7.8). Experience will soon tell you how many readings to take. Remember that if the jug fills quickly, use a bucket, and if that fills rapidly then you should be using a thin-plate weir as described below.

7.4 Stream-flow measurement

The flows of streams and rivers are either calculated from measurements

Location: Hogshead spring
Date: 22/10/86 (10.30 am)

Test number	Time to fill 1 litre jug (sec)
1	4.95
2	6.04
3	5.55
4	5.73
5	5.48

Average = $\dfrac{27.75}{5}$ = 5.55 seconds

∴ flow = 0.18 litres/sec

= 15.55 m³/d

Fig. 7.8 Field record of spring-flow measurements. The time taken for the jug to fill to the 1-litre mark is measured using a digital stop-watch. Measurements were repeated 5 times because of the apparent discrepancy between the first two readings. The average time is calculated and the reciprocal taken to give the average spring-flow in litres/second.

of water velocities and the channel cross-sectional area, or by installing a weir. If you are unable to obtain flow records for the stream that you are studying, from the local water authority or similar organization which measures river flow, you will have to take them for yourself, using the methods described here. It is always a good idea to check with the local water authority, as even if no suitable records are available they will probably be able to offer advice to help you make your own measurements. The methods which are described in the following sections are not suitable for floods or large rivers,

but nevertheless will meet the needs of most hydrogeologists.

7.4.1   Velocity–area method

Flow velocities are best measured using a current meter. These are devices which consist of an impeller which, when placed into a stream, rotates in proportion to the speed of the water passing by it. There are two main types of current meter, as illustrated in Fig. 7.9. Generally, because current-meters are expensive to buy and maintain, they will only be available to hydrogeologists working in organizations like water authorities, university departments and large consulting firms.

Current meters are used by suspending them in the stream, pointing in an upstream direction. For deep streams this must be done from a bridge or a specially constructed cableway. In most cases where you will need to take these measurements, you will probably be able to stand in the stream, wearing thigh-waders. In such cases the current meter is attached to a special pole, graduated in centimetres, which allows the instrument to be positioned into the river to a known depth. The rate at which the impeller rotates is recorded by an electronic counter suspended around the operator's neck and attached to the current-meter by a thin cable.

Ideally, you should choose a straight stretch of stream with a fairly constant depth in which to carry out the gauging. Choose a place where the

101

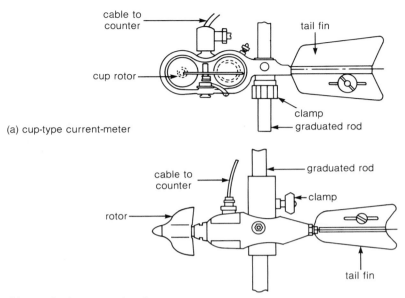

cable to counter

tail fin

cup rotor

clamp
graduated rod

(a) cup-type current-meter

cable to counter

graduated rod

clamp

rotor

tail fin

(b) propeller-type current-meter

Fig. 7.9 This figure shows two types of current-meter in common use. The cup-type (a) has an impeller made up of six small cups which rotate on a horizontal wheel. The rotor type (b) has an impeller which looks like the propeller on an outboard motor. In both types the rate of rotation of the impeller is recorded by an electrically operated counter and is converted to a velocity using a calibration chart. These instruments should be periodically recalibrated by a specialist firm.

flow is not impeded by cobbles, boulders, bridge supports, etc. Figure 7.10 shows the steps to take when making a current meter gauging. Firstly, stretch a surveyor's tape across the stream at right angles to the direction of flow. You will use this to identify each station where you take a velocity reading. These should be taken at regular intervals across the stream, spaced so that there are between 10 and 20 stations. Choose equal spacings of an easy size for the calculation, such as a metre or half a metre.

In theory, the average velocity occurs at 0.6 of the depth of the stream measured downward from the surface, and this is the position that all of the textbooks tell you to take your reading. For most practical purposes, however, it is as accurate to take the measurement at half the depth and then use a correction factor when the flow has been finally calculated (see Fig. 7.11). After all, when you are standing in a river, it is much easier to halve a depth in your head than to multiply it by 0.6!

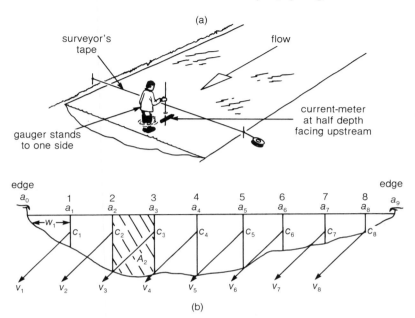

Fig. 7.10 Stream flow is often measured by means of a current-meter gauging, where stream-flow velocities are measured at regular intervals across the stream as shown in (b) The gauger faces upstream and stands to one side of the instrument so that turbulence around his legs does not affect the readings. Figs. 7.11 and 12 give details of how to record field measurements and also how to calculate the flow. (Chapter 10 includes safety precautions which must be observed when taking this type of field measurement.)

At each station first measure the depth using the graduated pole and set the current meter at half this value. Place the bottom of the rod on the stream bed, so that it stands vertically and at the precise position on the surveyor's tape. The current meter should point straight upstream and you should stand downstream and slightly to one side so that the measurement is not affected as the water flows round your legs. Allow the rotation of the impeller to settle down for a few seconds before start-

ing to count. Most modern instruments have a timing device in the counter, otherwise use a stop-watch and record the number of counts over a one-minute period.

The number of the station, the distance from the bank, the depth of water and the number of rotations should be recorded in your notebook as each measurement is taken. The calibration information for the current meter is used to convert the number of rotations into a velocity. The average velocity for each segment

(n) Section number	(a) Tape reading (m)	(w) Section width (m)	(c) Depth reading (m)	(d) Mean depth of section (m)	(A) Area section (m²)	(r) Current-meter reading at 0·5 depth — Field (rpm)	(x) Velocity (from tables) (m/s)	(v) Mean velocity of section (m/s)	(Q) Flow for section (m³/s)
Edge	a_0		Zero	$d_1 = 2 \times \dfrac{c_1}{3}$	$A_1 = w_1 \times d_1$	Zero	Zero	$v_1 = 2 \times \dfrac{x_1}{3}$	$Q_1 = A_1 \times v_1$
		$w_1 = a_1 - a_0$							
1	a_1		c_1			r_1	x_1		
		$w_2 = a_2 - a_1$		$d_2 = \dfrac{c_1 + c_2}{2}$	$A_2 = w_2 \times d_2$			$v_2 = \dfrac{x_1 + x_2}{2}$	$Q_2 = A_2 \times v_2$
2			c_2			r_2	x_2		
		$w_3 = a_3 - a_2$		$d_3 = \dfrac{c_2 + c_3}{2}$	$A_3 = w_3 \times d_3$			$v_3 = \dfrac{x_2 + x_3}{2}$	$Q_3 = A_3 \times v_3$
3	a_3		c_3			r_3	x_3		
		$w_4 = a_4 - a_3$		$d_4 = \dfrac{c_3 + c_4}{2}$	$A_4 = w_4 \times d_4$			$v_3 = \dfrac{x_3 + x_4}{2}$	$Q_4 = A_4 \times v_4$
etc.	etc.		etc.			etc.	etc.		
		etc.		etc.	etc.			etc.	etc.
Edge	a_9	$w_9 = a_9 - a_8$	Zero	$d_9 = 2 \times \dfrac{c_9}{3}$	$A_9 = w_9 \times d_9$	Zero	Zero	$v_9 = 2 \times \dfrac{x_8}{3}$	$Q_9 = A_9 \times v_9$

Calculation of stream flow (see Fig 7.10)
(1) Enter field readings n, a, c and r and obtain x from tables supplied with the current meter.
(2) Calculate w, d, A, v and Q.
(3) Sum values of Q.
(4) Correction for 0.5 depth (see section 7.4.1).
Stream flow $= 0.95 \times Q$ m³/s
$= 950 \times Q$ litres/sec

Total Flow $= Q$

Fig. 7.11 This diagram shows the field measurements to record during a stream-flow gauging, and how to use them to calculate the flow. All the symbols have the same meaning as Fig. 7.10. Take care to record all depths and distances in *metres*, not centimetres. Convert the number of revolutions of the current-meter into a velocity, by using the tables supplied with the instrument.

Date: 11TH JULY 1986				Stream: Farbeck: (U/S Farby Road Bridge)					
Time: START: 12·45		FINISH: 13·05							
n	*a*	*w*	*c*	*d*	*A*	*r*	*x*	*v*	*Q*
Section number	Tape reading (m)	Section width (m)	Depth reading (m)	Mean depth of section (m)	Area of section (m2)	Current-meter reading at 0.5 depth: Field revs in 50 secs	Current-meter reading at 0.5 depth: Velocity (m/s)	Mean velocity of section (m/s)	Flow for section (m³/s)
1	0.75		edge			—			
		0.25		0.167	* 0.0418			—	—
2	1.0		0.25			no detectable flow	—		
		0.5		0.280	0.1400			0.229	0.0416
3	1.5		0.31			278	0.343		
		0.5		0.335	0.1675			0.354	0.0593
4	2.0		0.36			298	0.365		
		0.5		0.365	0.1825			0.330	0.0602
5	2.5		0.37			235	0.295		
		0.5		0.395	0.1975			0.282	0.0557
6	3.0		0.42			211	0.269		
		0.5		0.415	0.2075			0.284	0.0589
7	3.5		0.41			237	0.298		
		0.5		0.385	0.1925			0.266	0.0512
8	4.0		0.36			180	0.234		
		0.5		0.405	0.2025			0.236	0.0478
9	4.5		0.45			183	0.238		
		0.5		0.435	0.2175			0.223	0.0485
10	5.0		0.42			156	0.208		
		0.5		0.435	0.2175			0.169	0.0368
11	5.5		0.45			87	0.131		
		0.5		0.425	0.2125			0.152	0.0323
12	6.0		0.40			124	0.172		
		0.5		0.340	0.1700			0.151	0.0257
13	6.5		0.28			86	0.130		
		0.5		0.440	0.2200			0.087	0.0342
14	7.0		0.16			no detectable flow	—		
		0.4		0.107	0.0428			—	
15	7.4		edge			—			
					*			Total flow	0.5522

* Note: As there was no detectable flow at 2 and 14 section areas have been added.

Correct for 0·5 depth
flow = 0·5522 × 0·95
= 0·525 m³/s or 525 litres/sec.

Fig. 7.12 This example shows how field data have been used to calculate a stream flow using the method given in Fig. 7.11.

is calculated and multiplied by the cross-sectional area to give the flow for each segment. The sum of the segment flows is the stream flow. Figure 7.11 shows how these field measurements are recorded and used to calculate the stream flow, with an example being given in Fig. 7.12.

It is possible to make a rough and ready estimate of the flow in a stream, using a crude form of the velocity–area method. This technique is useful when you are making an initial reconnaissance of the study area and want to get some idea of the general order of stream flows.

First, choose a length of stream which runs straight for 4 or 5 metres. Use a steel tape to measure both the width and depth of the channel at right angles to the direction of the flow. Multiply these values to get the cross-sectional area of the stream. If the stream bed is irregular, measure the depth at various distances from the bank and calculate an average depth.

The water velocity can be estimated by timing a piece of stick, or other suitable small float, over a measured distance of a metre or two. If the water is flowing rapidly it may be prudent to increase this distance to 5 or even 10 metres. You should take several readings with the float being placed at different distances across the stream, and then take the average.

The flow is calculated by multiplying the velocity measured in metres per second, by the cross-sectional area measured in square metres. This will give the answer in cubic metres per

second, and to convert to litres per second simply multiply by 1 000. This method will tend to overestimate flows, as the surface velocity is significantly greater than the average velocity. *It is important, therefore, to correct the estimated value by multiplying it by 0.75.*

7.4.2 Thin-plate weirs

The most accurate method of measuring the flow of a stream is to use a thin-plate weir. All weirs work by restricting the size of the stream channel. This causes the water to pile up on the upstream side, before passing over the weir as a jet. The rate at which water flows over the weir depends on the height of water above the weir on the upstream side. It is relatively easy to record these water levels and from them assess the flow, using standard tables.

There are two main types of thin-plate weir, a v-notch weir and a rectangular weir. The dimensions and method of operation of these weirs are described in the standards used by countries, e.g. in the UK, British Standard 3680 Part 4A. The specification given in such standards will ensure accurate flow measurements to within one per cent, although it is rarely necessary for such an accuracy in most groundwater studies.

You do not need to buy a special weir plate, as they are quite easy to make. Provided that you follow the general guidelines given here for their construction and installation, it is

106

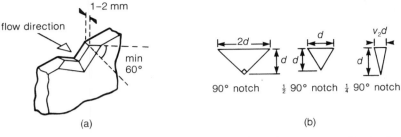

Fig. 7.13 It is important that a v-notch weir is made to the correct specification for accurate measurements to be achieved. The general arrangements shown in (a) apply to the three types of v-notch shown in (b).

possible to achieve flow measurements with an accuracy better than ±10%. Sheet metal or plywood can be used to make the weir plate. It is important to cut the angles as accurately as possible, to make sure that all edges are sharp and straight and that the upstream face is smooth. Figure 7.13 shows the general features of v-notch weir plates. The weir should have a lip of between 1 mm and 2 mm and the downstream face should slope away from the lip at an angle of at least 60° in the case of a v-notch weir. For rectangular weirs this angle must be at least 45°. Flows of up to 60 or 70 litres per second can be measured with a v-notch weir and higher flows can be measured with a rectangular weir of appropriate width.

There are three types of v-notch weir, each with a slightly different shaped notch. These are a 90° notch, which has its width across the top equal to twice the depth, a ½ 90° notch with the width across the top equal to the depth, and a ¼ 90° notch which has its width across the top equal to half the depth. If these

dimensions are used, it is a relatively simple job to construct the weir by cutting an appropriate isosceles triangle from the plate. The angles of the ½ 90° v-notch weirs are not 45° and 22.5° as might be expected. These weirs get their name from the fact that they measure half and one-quarter the amount as a 90° v-notch.

7.4.3 Installation and operation of thin-plate weirs

When a weir plate is installed in a stream channel it is important to select a straight section of channel at least three metres long. Use a builder's level to make sure that the plate is upright and ensure that it is at right angles to the direction of flow. It is vital to prevent water from leaking around the edges or underneath the weir. The edges can be sealed with clay, which often is best pushed in with the fingers. The level of the crest of a rectangular weir or the apex of a v-notch weir should be set at a height above the stream bed that will ensure

107

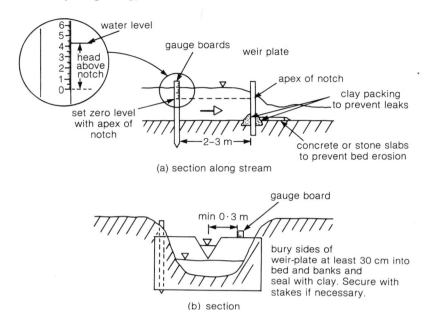

Fig. 7.14 Divert the stream or at least reduce the flow using sand-bags, while you install a weir-plate. Dig out a small trench in the bed and banks at right angles to the flow, to take the plate. This must be done quickly to reduce the problem of the trench collapsing. Drive stakes into the bank to support the weir-plate if needed. It is vital to make a good seal in both bed and banks, otherwise there will be leakage causing erosion followed by eventual collapse of the structure. Use clay to make this seal, either found locally or imported. Commercial bentonite may be used if clay is not easily obtained. Prevent bed erosion, which will also cause collapse, by using concrete slabs or stones. Take care when removing the sand-bags to restore the flow, so as not to cause a surge of water which may wash out the weir.

an adequate fall on the downstream side. At maximum flows this must be at least 75 mm above the water level in the channel on the downstream side (see Fig. 7.14). Installing these weir plates is not an easy job, often requires two people, and is likely to make you both wet and dirty. Do not take short cuts as poor workmanship will mean that the weir will be washed out in a

few days or even less. As weirs work by controlling the level of the water surface on the upstream side, the flow of water over a weir is related to the depth of water above the weir crest. If this depth is known, the flow can be looked up in the appropriate tables. It is important for the level to be measured at the correct distance upstream of the weir-plate, as shown in

Table 7.1   Flow over thin-plate weirs in litres/second

Head in cm	v-notch weirs			Rectangular weirs			
	¼ 90°	½ 90°	90°	0.6 m width	1.0 m width	1.3 m width	1.6 m width
1	0.005	0.01	0.02	0.8	1.3	1.9	2.1
2	0.02	0.04	0.1	3.1	5.2	6.7	8.2
3	0.05	0.1	0.2	5.6	9.4	12	15
4	0.1	0.2	0.5	8.7	15	19	23
5	0.2	0.4	0.8	12	20	26	32
6	0.3	0.6	1.3	16	27	35	43
7	0.5	0.9	1.8	20	33	44	54
8	0.7	1.3	2.6	24	41	53	66
9	0.9	1.7	3.4	29	49	63	78
10	1.2	2.2	4.4	34	57	74	91
11	1.5	2.8	5.6	39	66	85	105
12	1.8	3.5	7.0	44	75	97	120
13	2.2	4.3	8.5	50	84	110	135
14	2.7	5.2	10	56	94	122	151
15	3.1	6.1	12	62	104	136	167
16	3.7	7.2	14	68	114	150	185
17	4.3	8.4	16	74	125	163	200
18	5.0	9.6	19	81	136	178	220
19	5.6	11	22	87	148	193	240
20	6.4	12	25	94	159	210	260
21	7.2	14	28	101	171	225	275
22	8.1	16	31	108	183	240	295
23	8.5	17	35	116	196	255	315
24	10	20	39	123	210	275	335
25	11	22	43	131	220	290	360
26	12	24	48	139	235	310	380
27	13	26	52	147	245	325	400
28	15	29	57	154	260	340	420
29	16	31	63	162	275	360	445
30	17	34	68	170	290	380	470

Fig. 7.14. The relationship between water level and flows over various types of thin-plate weirs is given in Table 7.1.

To take the measurements, install a gauge-board on a vertical post at the appropriate upstream distance. It should be positioned to one side of the channel rather than in the centre, and a builder's level should be used to ensure that the post is vertical. The gauge-board should be accurately marked off in centimetre graduations. A wooden ruler can be used as a ready-made gauge-board. You can paint alternate marks in a distinctive colour to help to read the water level to the nearest centimetre. The zero of the gauge-board needs to be at the same level as the crest of the rectangular weir or the apex of the v-notch.

The biggest drawback to this method of measuring stream flows is that the weir will tend to silt up on the upstream side. It is vital to keep it cleaned out so that the water on the upstream side is deep enough for the weir to work. For a rectangular weir, the minimum depth of water required is 60 cm. A v-notch which is 30 cm high will need a depth of 45 cm, and a notch which is 15 cm high will require 30 cm of water. It is important to clean the section out *before* the flow measurement is taken. In addition, remove any debris such as twigs, branches, waterweeds or polythene bags from the weir. Do not take the reading until stable conditions have been re-established. This may take several minutes, so be patient. Normally one reading would be taken each day and be used to build up a record of flows as shown in Fig. 7.15.

7.5 Stage-discharge relationships

In some circumstances it will not be possible to install a thin-plate weir, but stream flows can be measured by relating stream levels as measured on a gauge-board to a series of current meter gaugings. Try to select a straight length of stream where flows are controlled by a natural feature of the stream channel such as an outcrop of rock. If this is not possible, an artificial control can be installed which need only be something as small as a length of two inch diameter pipe laid across the stream-bed at right angles to the flow, and held in place by stakes driven into the bed and banks. The gauge-board should be installed four or five metres upstream of the control, on a wooden stake which is set vertically into the streambed. Current meter gaugings are made as described above but with the gauge-board being read before and after the gauging is made. In this way the relationship between stream level (stage) and flows (discharge) can be built up. Eventually, sufficient gaugings will have been made to enable a stage–discharge graph to be drawn (Fig. 7.16). This can be used to interpolate flows from gauge-board readings, without a current-meter measurement being carried out. For a continuous record, install a water level recorder which is set to the same datum as the gauge-board. The recorder will need to be in a box to protect it from the elements and from vandals. The float should be installed in a stilling well. This need only consist of a large diameter pipe set vertically at the river bank with adequate holes drilled in it to enable the

Day	October '86		November '86	
	Head above notch (cm)	Flow m³/d	Head above notch	Flow m³/d
1	—	—	5	70
2	—	—	6	112
3	—	—	6	112
4	—	—	7	155
5	—	—	8	225
6	—	—	10	380
7	6	112	12	605
8	7	155	12	605
9	9	294	11	484
10	7	155	10	380
11	5	70	9	294
12	8	225	8	225
13	10	380	8	225
14	7	155	7	155
15	6	112	—	(133)
16	5	70	6	112
17	6	112	5	70
18	5	70	7	155·
19	—	(70)	10	380
20	5	70	—	(380)
21	4	43	10	380
22	4	43	9	294
23	4	43	8	225
24	5	70	7	155
25	6	112	6	112
26	—	(168)	6	112
27	8	225	5	70
28	8	225	8	225
29	6	112	10	380
30	5	70	12	605
31	5	70	12	605

Fig. 7.15 Example of a stream-flow record. Flows were measured with a 90° v-notch weir, with the head measured at the same time each day. This gave an instantaneous flow which has been taken as the daily average. The record starts on 7 October. Missed readings have been interpolated and are shown in brackets.

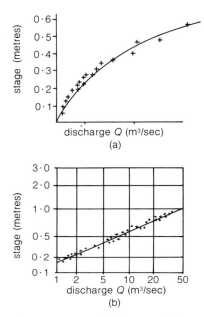

(a)

(b)

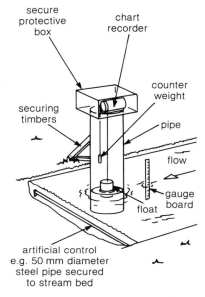

Fig. 7.16 A stage–discharge graph shows the relationship between stream flow and water level at a particular point. Simple relationships can be seen from a plot on a natural scale (a), but more complex relationships are best understood if a log–log graph is plotted, (b).

Fig. 7.17 A continuous record of stream level is made by the temporary installation of a recorder. In this example, a stilling well has been made using a vertical pipe secured to the bank with timbers. An artificial bed control has been provided by securing a 50 mm diameter pipe to the stream bed at right angles to the flow.

water level in the pipe to be the same as that of the stream. This arrangement protects the float from damage and removes small fluctuations in water level caused by waves, etc. Figure 7.17 shows an example of this type of installation.

7.6 Choosing the best method

The method you choose to measure the flow of springs and streams will depend on local circumstances. It is vital to select one which will cope with the range of flows you need for your investigation. In many groundwater studies, the low flows, i.e. those which occur during dry weather, are most important, and so you should give some thought as to the most appropriate method of field measurement to use before a start is made to install a thin-plate weir or a clay dam and pipe. Table 7.2 will

Table 7.2 Maximum flows for accurate measurements with different methods

Method of measurement	Maximum flow for accurate measurement (litres/second)
Jug and stop-watch	
1-litre jug	0.25
10-litre bucket	3
Weir-plates	
¼ 90° v-notch	17
½ 90° v-notch	34
90° v-notch	68
Rectangular weirs	
0.6 m width	170
1.0 m width	290
1.3 m width	380
1.6 m width	470

give you an idea of the maximum flows which can be measured accurately with each method.

It is also important to examine the stream-channel dimensions and consider which system can be installed from a construction point of view. A wide, shallow channel with a small downstream slope, for example, may be unsuitable for using a thin-plate weir, and you may have to resort to the jug and stop-watch method after building the sort of installation shown in Fig. 7.7. One word of warning before you get out your spade and start to dam up a stream. In Britain and some other countries it is necessary to obtain the permission of the water authority or local authority to install equipment in a water course. Quite often the formal aspects will be minimized and probably you will only have to let the local office know your proposals. This may well have hidden bonuses because the resident hydrologist will probably be very pleased to provide advice on the best way of tackling the problem.

The frequency with which you need to take measurements will depend upon the type of study being carried out. Daily readings may be needed, but in many instances two or three readings a week will do. If daily readings or even more frequent ones are required, it may be better to install a chart recorder as described above, so that a continuous record of levels is obtained from which flows can be calculated.

7.7   Processing flow data

Whatever the frequency of measurements, the best way to see how they change is to plot them against time. This is an excellent way of comparing one measurement with another and, if you include rainfall readings, it is possible to tell how quickly flows respond to rainfall events. Figure 7.18 shows an example of this type of graph. Prepare the data by converting flows to the same units, and then select a time-scale large enough for all measurements to be plotted against the time they were taken. This is not always important, but it can be useful when attempting to separate the

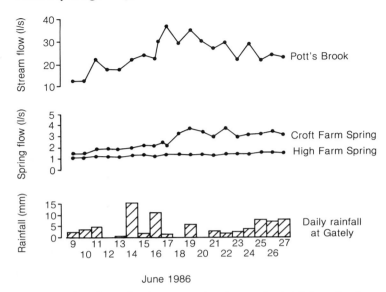

Fig. 7.18 The relationship between spring flow and rainfall is easily discovered if records are plotted on the same time-scale. In this example, the rapid response of Pott's Brook to rainfall is obvious, as is the different response of the two springs. The Croft Farm Spring flow increases with rainfall, suggesting that it is the discharge from a shallow aquifer or even a land-drain. The flow of the High Farm Spring changes very little, indicating that it is the discharge from a separate, deeper aquifer. (North West Water data.)

effects of a nearby pumping borehole from those caused by rainfall, for example.

You should expect a short time-lag of only a few hours, before flows of streams and rivers start to increase flow after heavy rain. The precise length of this response depends upon many factors, such as catchment size, slope and geology, the amount of rainfall over the recent past few days, the intensity of the rain and the direction in which the storm was travelling across the catchment. Groundwater levels, on the other

hand, respond much more slowly. Water levels in deep aquifers may only show general seasonal trends, while those in shallow aquifers may show a rise within a day or so of heavy rain. Variation in spring flows generally follows the same pattern. Those which drain deep aquifers change very little throughout the year, and you are unlikely to see the effect of an individual rainfall event. The flow of springs fed by shallow aquifers varies much more and this factor can be used to identify each type of spring. Figure 7.18 is based on field data, and shows how marked these differences can be.

If you collect sufficient stream-flow data, you can examine it using a technique called 'baseflow separation'. This is one method of hydrograph analysis, a subject which is generally studied in the office rather than in the field. These techniques use the fact that the total flow of a stream is made up of a number of components, each of which behaves differently. The two major components are surface runoff and groundwater drainage. The latter is made up of spring flows and direct discharges into the stream, and is referred to as *baseflow*.

One graphical technique is straightforward to use. It involves plotting flow data on semi-logarithmic graph paper, with time along the bottom on a natural scale (Fig. 7.19). On this graph, the decline in groundwater flows (or baseflow recession) approximates to a straight line. Once the data are plotted, examine the slope of the graph after single-peak storms. The decline in total stream flow will follow the baseflow recession line at the later stages of this reduction in flow. Compare a number of single storm events. This is important, as the baseflow recession will be a constant and so the various straight lines should be parallel. Once the slope of this line has been identified, you can use it to draw in the baseflow in those parts of your records where the straight line is not so obvious. In this way it is possible to separate the groundwater component from the other elements of stream flow.

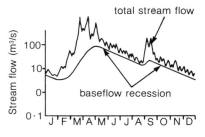

Baseflow recession curve

Fig. 7.19 A baseflow recession graph can be constructed from a semi-logarithmic plot of stream flow with time. The rate of the baseflow recession is constant, which produces a series of parallel lines as shown.

Spring flows can be examined in the same sort of way, but this is only likely to be worthwhile where the flow varies a great deal. As many 'springs' may really be land-drains or at least have a surface water component, it can be important to identify the groundwater element. This will help you to decide which aquifer unit a spring is draining, and will indicate whether or not it is likely to dry up. Both of these facts are fundamental to a full understanding of the workings of the groundwater system in your study area.

7.8   Using chemical data

Water chemistry can also provide a valuable clue as to the nature of springs. Samples of water should be taken from all wells and springs for chemical analysis, as a matter of

115

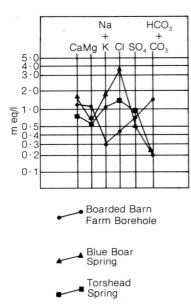

Boarded Barn
Farm Borehole

Blue Boar
Spring

Torshead
Spring

Fig. 7.20 Schoeller semi-logarithmic plot of groundwater chemistry data. The concentration of each ion, expressed in milli-equivalents, is plotted as shown. The relationship between two chemical constituents in different water samples is shown by the slope of the straight lines connecting these constituents. Parallel lines indicate identical relationships. In this example, the chemistry of the two springs is very similar, whereas the water from the borehole is quite different, indicating that it penetrates an aquifer separate from the one supporting the springs. (North West Water data.)

routine. This will enable you to compare the chemistry of different spring waters with that of groundwater taken from various wells. The chemistry can be used to 'fingerprint' these waters and thereby separate surface water from groundwater and also

Table 7.3 Conversion factors for expressing the concentration of selected ions in milli-equivalents

Ion	Factor
Calcium (Ca)	20.04
Magnesium (Mg)	12.16
Potassium (K)	39.10
Sodium (Na)	23.00
Chloride (Cl)	35.46
Sulphate (SO_4)	48.04
Nitrate (NO_3)	62.01
Bicarbonate (HCO_3)	61.01
Carbonate (CO_3)	30.01

The milli-equivalent value is calculated by *dividing* the concentration of each ion in milligrams per litre (or ppm) by the factor shown.
Note: nitrate as mg/l nitrogen = nitrate as mg/l nitrate × 0.2258; bicarbonate as mg/l HCO_3 = alkalinity in mg/l $CaCO_3$ × 1.22.

differentiate groundwaters from separate aquifer units. This is described more fully in the companion Handbook on groundwater quality and chemistry.

A useful graphical technique was developed by a hydrogeologist called Schoeller, after whom the method is called. It involves plotting the concentrations of the six principal chemical components of Ca, Mg, Na+K, Cl, SO_4 and HCO_3+CO_3 on semi-logarithmic graph-paper. These values are plotted at arbitrary but evenly spaced distances along the arithmetic scale, as shown in Fig. 7.20. The concentrations of each ion are plotted in milli-equivalents per litre along the logarithmic scale. The milli-equivalent value is calculated by dividing the concentration of each ion in milligrams by its atomic weight — which has been divided in its turn by the valency. Where the concentrations of two ions are plotted together as $Na+K$ and HCO_3+CO_3, calculate the milli-equivalent value of each before adding them together. Table 7.3 will help make these calculations easy.

Another graphical method of examining groundwater chemistry data was developed by Hill (1940) and Piper (1944), and is generally referred to as the *trilinear graph* (Fig. 7.21). This method has a greater potential than the Schoeller diagram, as it will accommodate results from more analyses and can be used to show the effect of mixing two waters.

7.9 Using artificial tracers

The primary use of tracers is to demonstrate the underground inter-connection between sink-holes where water disappears underground, and springs – mainly in karstic limestone

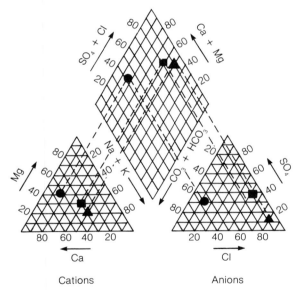

Fig. 7.21 Trilinear plot of a type proposed by Piper (1944). The concentrations of cations and anions are plotted as the percentage of the total of each when expressed in milli-equivalents. The grouping on the central diamond-shaped field shows up waters which have a similar chemical composition. The data used in this diagram are the same as those used in Fig. 7.20. (North West Water data.)

117

aquifers – but there are some applications in aquifers with other lithologies. Tracing involves adding a material to water at the point where it disappears below ground, and then monitoring all possible outflow points for signs of the tracer. This enables flow paths to be identified and travel times to be calculated. Besides its important role in karst studies, the technique is also used in quantifying resources and to locate the source of a pollutant.

The tracing method works best in fissured aquifers, especially limestones. It has been used successfully in some intergranular aquifers, but it often fails because the tracer may be filtered out; the more intimate contact with the rock may encourage removal of the tracer by absorption or chemical reaction; or dispersion of the tracer by intergranular flow may cause it to be diluted below detectable concentrations (see section 5.2.3). A detailed guide to the use of tracers is given by Drew and Smith (1969). Figure 7.22 shows how tracers can be used to understand the complex interrelationships between sink-holes and springs. An ideal tracer should be detectable in minute concentrations, should not occur naturally in the tested waters, should not react chemically with or absorb into the aquifer material, should be safe in terms of health and the aquatic environment, and should be cheap and readily

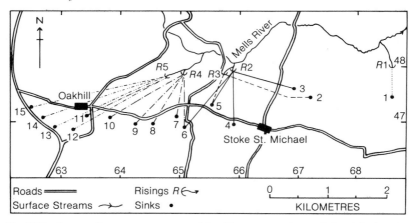

Fig. 7.22 *Lycopodium* spores were used in this tracer study carried out in karstic Carboniferous Limestone in the Mendip Hills, southwest England. The spores introduced at each sink or swallow-hole were identified with a distinctive dye. This method allowed up to six swallow-holes to be tested at a time. The field results allowed the relationship between 15 swallow-holes and five risings or springs to be established. It showed a complex fissure-flow geometry, where most risings are supported by several sinks and some sinks are connected to more than one rising. (Figure reproduced from Drew and Smith, 1969 by permission of Geobooks Ltd.)

available. The materials commonly used as tracers fall into five groups:

(i) *Dyes.* These give a distinctive colour to the water, using dye-stuff which is visible in low concentrations. Detection is either by direct observation or by concentrating the tracer on to hanks of cotton or absorption on to charcoal. Some of the dyes used are fluorescent and can be detected in low concentrations with a fluorimeter. Usually samples are taken at regular intervals and then examined in a laboratory. Commonly used dyes are fluorescein (orange colour and green fluorescence), pyranine concentrate (similar to fluorescein but more resistant to absorption), congo red (a red dye which may turn blue in acid waters) and rhodamine B (a red dye similar to fluorescein).

(ii) *Chemicals.* A concentrated solution of an inorganic salt is added to the water at the sink-hole. The most usual chemicals are soluble chloride and sulphate salts and sugars. The tracer is detected by frequent sampling followed by an appropriate chemical analysis. Often electrical conductivity variations are used to monitor changes in concentration, because conductivity changes significantly with increases in dissolved solids. These measurements can be taken in the laboratory or on-site, using a portable probe. If the probe is connected to a chart recorder or a data logger, a continuous record can be obtained which may be used to calculate travel time in detail.

(iii) *Mechanical.* A suspension of fine particulate matter is added to the water and detected either by visual observation or by capture on nets. *Lycopodium* (a species of club-moss) spores are the most commonly used material. Where tracers are to be added to several sink holes, the spores are dyed to identify their origin. This technique allows up to six sink holes to be tested at once. Other mechanical tracers, including small floats of polystyrene or polypropylene are also sometimes used, although Drew and Smith (1969) report that the only successful mechanical tracers are spores.

(iv) *Bacterial.* A distinctive bacterial culture is added to the water and detected by sampling and culture growth in the laboratory. This method is rarely used because of environmental objections and tedious sampling and laboratory procedures. Where a spring is contaminated by a bacterium, however, it may be possible to use this as a tracer to identify the likely source of the contaminant. Once this has been found, the connection should then be demonstrated using another tracer technique.

(v) *Radioactive.* This is really a variation of the chemical tracer

119

Table 7.4 Summary of the characteristics of groundwater tracer techniques

Techniques	Soluble in or unaffected by acid solution	Soluble in or unaffected by alkaline solution	Not absorbed in transit	Toxicity	Not objectionable	Not time-consuming	Inexpensive	Detectable at low concentrations	Flow rate similar to water	Not naturally present in the water	Without effect on the water	Several traces simultaneously possible	Continuous monitoring of rising unnecessary	High sensitivity	Simplicity	Provides subsidiary statistics
Dye tracers																
Fluorescein/direct observation	+	+	+	‡	—	—	+	‡	—	‡	‡	—	+	—	‡	—
Congo red	‡	+	+	‡	—	—	+	+	—	‡	‡	—	+	—	‡	—
Pyranine/direct observation	‡	‡	‡	‡	—	—	—	‡	—	‡	‡	—	+	+	‡	—
Rhodamine-B/direct observation	‡	‡	+	—	—	—	—	‡	—	‡	+	—	+	—	‡	—
Fluorescein/charcoal method	+	+	‡	‡	‡	‡	‡	‡	—	‡	‡	—	‡	—	‡	—
Pyranine/charcoal method	‡	‡	‡	‡	‡	‡	+	‡	—	‡	‡	—	‡	+	‡	—
Rhodamine-B/cotton hanks	‡	‡	+	—	‡	+	‡	‡	—	‡	+	—	‡	—	‡	—

Technique													
Malachite green/cotton hanks	++	++	—	++	++	+	+	—	++	+	++	‡	—
Durazol orange/nylon detectors	++	+	+	+	++	+	+	—	++	+	++	‡	—
Non-dye tracers													
Polypropylene floats/nets	++	++	++	++	++	‡	‡	—	++	—	‡	+	+
Lycopodium/plankton nets	++	++	++	++	++	—	—	++	++	‡	‡	—	++
Radioactive tracers (salts)	++	+	—	—	—	—	—	++	++	+	—	+	—
Bacterial culture	+	+	+	+	+	+	+	+	++	+	—	—	—
Potassium salts/flame photometer	++	++	++	+	+	+	+	+	++	+	—	+	—
Ammonium chloride/electrical conductivity	++	+	+	—	+	+	—	+	—	+	+	—	—
Inorganic salts/chemical analysis	++	+	+	+	+	+	+	+	+	+	—	+	—
Tritium	++	++	++	+	—	—	—	‡	‡	+	—	—	+

— poor   + fair   ++ good

(Adapted from Drew & Smith 1969 by permission of Geobooks Ltd.)

Field Hydrogeology

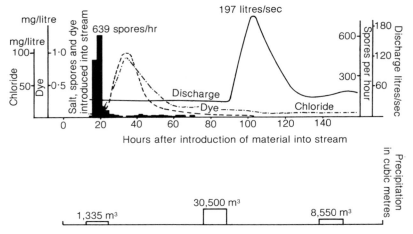

Fig. 7.23 The concentration of *Lycopodium* spores, fluorescein dye and sodium chloride solution are compared as they flow along a stream. Stream flow is also shown, as is catchment rainfall. The graph shows how each tracer has been dispersed between the injection point and monitoring station, and how the flow responds to rainfall. From this experiment it was concluded that spores are the best tracer for closely reflecting the maximum flow-rate of the water. (Figure reproduced from Drew and Smith, 1969 by permission of Geobooks Ltd.)

method. The commonly-used tracers are [131]iodine, [60]cobalt, [86]rubidium and tritium ([3]hydrogen). Detection is made using a Geiger counter, but these tracers are rarely used because of general objections to introducing radioactive materials into the environment, and also because of cost. Radioactive isotopes which occur naturally in groundwater, however, are used as tracers without any environmental objection. All radioactive tracers require the use of expensive analytical methods and careful sampling techniques. A detailed description of these methods can be found in Brown *et al.* (1983), Freeze and Cherry (1979) and Todd (1980).

The relative merits of commonly used tracers are summarized in Table 7.4. It is important to consider the attributes of particular tracers when deciding which to use, although other factors such as cost and availability may be equally important. Not all tracers move at the same velocity. Figure 7.23 compares three different tracers used in a karstic limestone and indicates that *Lycopodium* spores are the best in this groundwater environment. On the other hand, pyranine concentrate is likely to be the most successful tracer in an intergranular aquifer.

Before using any form of tracer, it is important to seek permission from the local authorities responsible for public health and water quality. Do not forget to carry out a survey to identify all the possible locations where the tracer may emerge. This is needed to set up the sampling points, but it is also required to ensure that co-operation is obtained from land-owners and to avoid water users being surprised by a sudden change in the colour of their stream.

Remember that if you fail to identify any tracer at the measuring point, it does not necessarily mean that there is no connection. Negative results are not conclusive. Tracer experiments should be used only as part of an overall study of ground-water flow.

8
Water balance

An important objective of most groundwater studies is to make a quantitative assessment of the groundwater resources. This can be considered in two ways the total volume of water stored in an aquifer and the long-term average recharge to the aquifer. The more significant figure in terms of groundwater resources is the long-term average recharge, and it is usual for this value to be regarded as the available resources. Because groundwater is a renewable resource, development should not cause permanent depletion of aquifer storage. It is inevitable that pumping from boreholes will cause some local lowering of water table levels in order that groundwater will be induced to flow towards the wells.

Once a new groundwater level equilibrium has been established, levels will stabilize and only fluctuate in response to annual variations in recharge, seasonal changes and other natural causes. If resources are over-developed and abstraction rates exceed the average recharge, ground-water levels will decline. This causes shallow wells and springs to dry up and increases the cost of pumping

from deeper wells and boreholes. Eventually, groundwater will no longer be available for abstraction. Before this happens, other serious problems may develop, particularly a deterioration in water quality caused by such factors as sea water intrusion into coastal aquifers or the up-coning of deep-seated mineralized ground-waters. These changes in ground-water quality are likely to mean that wells are abandoned even before the aquifer is pumped out. It follows, therefore, that resource management is essential and a knowledge of the annual average recharge is basic to this process.

In Chapter 5, we saw how it is possible to calculate the volume of water stored in an aquifer by using geological information to define the total volume of water-bearing rock and from a knowledge of the aquifer specific yield. This quantity is important as it represents the reservoir of groundwater available for abstraction. In aquifers where there is a large volume of water held in storage, this reservoir acts as a buffer, allowing abstraction to continue in drought years when there is little recharge.

8.1 Water balance

Groundwater recharge to an aquifer cannot be measured directly, but only inferred from other measurements. As groundwater is part of the hydrological cycle, measurement of other components of the cycle can be used to estimate the value of the resources, using a technique called a *water balance*. In this type of calculation it is assumed that all the water entering an area is equal to the water leaving an area, plus or minus any change in storage. This can be written more fully as in the equation given below. The various elements of this equation are listed in Table 8.1, which includes comments on methods of estimating each individual component.

The water-balance method involves identifying which components of inflow and outflow can take place within an area and then quantifying each one individually. The water-balance equation is now used to determine the groundwater recharge in the way that a mathematical equation is used to determine the value of a particular unknown variable. As all the components of the water-balance equation are estimated from field measurements and observations, the calculated values can be added to the inflows or outflows of the equation and the overall accuracy of each estimate can be assessed by examining how well each side of the equation balances.

$$
\begin{pmatrix}
\textit{Inflows} \\
\text{rainfall + recharge from} \\
\text{surface water + sea-water} \\
\text{intrusion + inflow from} \\
\text{other aquifers + leakage +} \\
\text{artificial recharge}
\end{pmatrix}
=
\begin{pmatrix}
\textit{Outflows} \\
\text{abstraction + spring flow +} \\
\text{base flow in rivers +} \\
\text{discharge to the sea + flows} \\
\text{to other aquifers +} \\
\text{evapotranspiration}
\end{pmatrix}
\pm
\begin{pmatrix}
\text{change in} \\
\text{aquifer} \\
\text{storage}
\end{pmatrix}
$$

Table 8.1 Components of the water-balance equation

Types of flow	*Comments on methods of estimation*
(a) Inflows	
1 *Rainfall*	This is usually the most significant recharge component and consists of that proportion of rainfall which percolates into an aquifer (see section 4.5). Some rainfall is lost as evapotranspiration or runoff. Estimates are based on rainfall and evapotranspiration data and the consideration of the geology and groundwater levels.

Table 8.1—*(continued)*

Types of flow	Comments on methods of estimation
2 *Recharge from surface water*	Where streams, rivers, lakes or ponds have a permeable bottom or sides, water can percolate into an aquifer when the groundwater levels are lower. This recharge is estimated using Darcy's law by consideration of the geology and the difference between surface water and groundwater levels. Look for the evidence of changes in groundwater chemistry which may be caused by percolating surface waters (section 7.8).
3 *Sea-water intrusion*	When the water levels in coastal aquifers are lowered by pumping, the potential exists for sea-water intrusion. Apply Darcy's law (section 5.1) to calculate inflows from information on groundwater levels and aquifer hydraulic conductivity. Look for chemical evidence to show that intrusion is taking place (section 7.8).
4 *Flow from other aquifers*	All aquifers which are adjacent to the study area should be examined as potential sources of recharge. Geological information, groundwater levels and chemical evidence will help to decide if flow is taking place across boundaries. Estimate inflow using Darcy's law.
5 *Leakage*	This is an artificial type of inflow caused by leakage from water-supply reservoirs, water pipes and sewers, resulting from damage or deterioration. Leakage is estimated by measuring inflows and outflows of the water-supply or sewer system. In some areas, excessive irrigation of crops, or even parks and gardens, can provide a significant quantity of extra recharge. This can only be quantified by direct observation.
6 *Artificial recharge*	In some aquifers, natural recharge is artificially supplemented by water being recharged through special lagoons or boreholes. This component of

Table 8.1—*(continued)*

Types of flow	Comments on methods of estimation
	recharge can be easily quantified from direct readings. Sewage disposal may sometimes also be another source of recharge water.

(b)   Outflows

1 *Abstraction*	Use metered records whenever possible, otherwise estimate from the pump capacity and hours run or from the water requirements given in section 9.1. A discussion with the well operator will be essential. Do not forget to include other groundwater pumping, such as dewatering mines and quarries or major civil-engineering works.
2 *Spring flows*	Groundwater discharge from springs can be assessed by measuring each one separately or from stream-flow measurements, which will include the flow from a number of springs.
3 *Baseflow*	The groundwater component of surface water flows can be estimated from stream flow records. (See section 7.7.)
4 *Discharges to the sea*	In coastal areas, groundwater may discharge directly into the sea. Sometimes this forms a spring-line between high and low water marks which may be identified from observation, temperature or conductivity measurements. Apply Darcy's law to estimate the quantities involved.
5 *Flow to other aquifers*	Where an aquifer extends beyond the study area, groundwater may flow out of the study area, with no surface indication. Use groundwater level information and flow-net analysis (see sections 6.6 and 6.7) to estimate quantities. Discharges to other aquifers may occur along boundaries, and these should be estimated in a similar manner. Use

Table 8.1—*(continued)*

Types of flow	Comments on methods of estimation
	geological information, groundwater level records and chemical information to decide if such flows can occur.
6 *Evapotranspiration*	In areas where the water table lies close to the ground surface, groundwater may be removed by plants taking water up through their roots and transpiring it out through their leaves. Generally, however, evapotranspiration removes water from the soil, thereby creating a deficit which is made good by the following precipitation. This process reduces the quantity of rain-water available for recharge. Evaporation also takes place from bare soil, with water flowing upwards under capillary forces to replace losses. In some areas, groundwater may be exposed in quarries or ponds. Here, evaporation losses will take place at the potential rate, equivalent to losses measured in an evaporation tank (section 7.2).
(c) **Changes in storage**	These are estimated from field measurements of groundwater levels and the aquifer specific yield or storage coefficient. Do not forget to take into account whether the aquifer is confined or unconfined (see section 5.5).

8.2 Rainfall recharge

Potential rainfall recharge is the total rainfall, minus the losses from evapotranspiration. The remaining water will either percolate into the ground or run off to surface water-courses where it can be measured as part of the stream flow gauging programme. Evapotranspiration is the most diffi-cult part of the hydrological cycle to quantify. Penman (1948) derived an empirical relationship between evaporation from open water and evaporation from bare soil and grass. Working in Southern England, he concluded that the evaporation rate from a freshly wetted bare soil is about 90% of that from an open water surface exposed to the same

Table 8.2 Summary of the results of an experiment carried out by Penman (1948) on the relationship between evaporation losses from open water and those from bare soil and grassland. The work was carried out at Rothamsted in southern England

Type of surface	Percentage of evaporation from open water
Open water	100
Bare soil	90
Grassland	
winter	60
summer	80
autumn/spring	70
whole year	75

Adapted from Penman (1948).

weather conditions. His general conclusions are given in Table 8.2. Measurements of evaporation losses from an evaporation tank (see section 7.2) can be used to calculate field evapotranspiration losses using these factors.

Other workers have produced alternative methods of calculating evapotranspirational losses. Thornthwaite (1948) used an empirical approach to calculate evapotranspiration losses from grassland in the eastern USA, based on mean monthly temperature measurements. This is a widely used method, but it strictly only applies in areas with a similar climate to the area where it was developed, and can produce significant errors if used elsewhere. Penman (1950*a*, *b*) developed a more detailed approach using the concept of *soil moisture deficit*. This method allows for the fact that vegetation continues to transpire water from the soil during periods without rain, which creates a 'deficit' that takes up part of the following rainfall, thereby reducing recharge. Penman's methods have been further developed by Grindley (1969, 1970). These techniques are beyond the scope of this Handbook, and the reader is referred to the appropriate reference or general textbook.

9
Special groundwater investigations

Groundwater investigations are carried out for a wide variety of reasons. Some are concerned with water supplies – either the development of new ones, or examining the reasons why a supply has failed or has become polluted; others are to assess how waste can be disposed of safely without causing pollution of either surface waters or groundwaters. A sound understanding of local hydrogeological conditions, obtained by following the techniques set out in this Handbook, is the basis of all groundwater problem solving. Complex problems may need more sophisticated techniques, details of which are to be found elsewhere, but often the hydrogeologist needs to have a general understanding of other disciplines such as water engineering or even plumbing, and an ability to apply common sense. This chapter aims to describe the main considerations to bear in mind during a wide variety of different groundwater investigations. General advice is given on which of the techniques described in earlier chapters should be used to answer specific problems. In all cases a desk study should be carried out to identify the main aquifers and obtain a general understanding of groundwater flow direction.

9.1 New groundwater supplies

An early consideration in planning a new water-supply development is exactly how much water is needed. The actual quantities required for survival are small, about 5 litres per person per day for drinking and cooking, although 20–40 litres per person per day are typical of the total water use at subsistence level in less developed countries. In developed countries, 150–200 litres per person per day are used. The extra water is needed for water-borne sewage systems and for washing machines, dishwashers, car washing, etc. These figures do not take into account the agricultural requirements for stock watering and crop irrigation, or the needs of industry. Table 9.1 gives the water requirements for many purposes and will help you to estimate the water needs for any scheme where you are involved in finding adequate water resources. It is important to consult with the people who will be using the

Table 9.1 Average water requirements for various domestic purposes, agricultural needs and manufacturing processes

Use of product	Quantity of water needed*
Domestic use	
Drinking (per person per day)	2–3 litres
Washing-up (per time)	2–4 litres
Flushing lavatory (per flush)	12–20 litres
Bath	130–170 litres
Shower (per minute)	20 litres
Washing-machine (per load)	130 litres
Watering garden for one hour	1300 litres
Animals – Daily requirements	
Cow (milk producer including dairy use)	150 litres
Cow	50 litres
Horse	50 litres
Pig	15 litres
Sheep	7–8 litres
Poultry (per 100 birds)	25 litres
Food growing	
1 tonne wheat	1000 m^3
1 tonne rice	4500 m^3
1 tonne sugar	1000 m^3
1 tonne potatoes	550 m^3
Manufacturing	
1 tonne beer	$6–10 \text{ m}^3$
1 tonne bricks	$1–2 \text{ m}^3$
1 tonne steel	250 m^3
1 tonne aluminium	1500 m^3
1 tonne fertilizer	600 m^3
1 tonne refined crude oil	15 m^3
1 tonne synthetic rubber	3000 m^3

* Note that 1 m^3 ($=10^3$ litres) of water weighs 1 tonne.

Adapted from S238 by permission of the Open University.

water, to ensure that their exact needs are identified on an hourly, daily and annual basis. This breakdown is necessary because water is often only needed for a few hours each day, and not always for every day of the year. This is particularly true of water supplies to factories and farms. Try to identify the maximum quantities needed, but resist the temptation to overestimate, as this could result in unnecessary expenditure by making the source too big. Where water resources are limited, an overestimate of needs could result in expensive alternatives being developed, perhaps involving pipelines many kilometres long. Unnecessarily expensive water supplies may even result in a new factory not being opened or an existing one closed down, with all the social and economic consequences that will have.

9.1.1 Spring supplies

Some smaller groundwater supplies are obtained from springs. It is important to assess the reliability of all the possible springs which may be used; that is, compare the seasonal variation in flow to the required supply quantities. Establish a regular monitoring programme as set out in Chapter 7. Take a sample of water and have it chemically analysed and checked for bacterial content. This will be needed to see if it is fit for drinking or other uses. By comparison with samples from local wells, you will be able to decide whether the

spring is fed from a deep aquifer or is shallow and has a significant surface-water content. This will also be indicated by the change in flow in response to rainfall, best decided from a graph showing variations in flows and rainfall in daily totals.

Measures should be taken to see that a spring supply is protected from pollution by constructing a catch-pit (see Fig. 9.1). Pollution problems should be anticipated from sewage disposal, the use of agricultural chemicals such as fertilizers and pesticides and the disposal of agricultural waste such as silage liquor (produced from grass fermented to form cattle feed) or spent sheep dip. A knowledge of both the local hydrogeology and agriculture practices is needed here. Landfill sites (rubbish tips) are another potential source of pollution which you should be on the look out for when assessing a spring as a water supply.

In fissured aquifers, especially karstic limestones, or where there is a steep topography, groundwater flow is rapid and pollution of spring supplies can happen quickly after noxious materials are spilt or deliberately spread on to land. It is prudent to carry out a thorough hydrogeological investigation, including tracer studies, to establish the vulnerability of the supply to pollution hazards.

9.1.2 The safe yield of wells and boreholes

The safe yield of a new well or borehole is usually uppermost in the

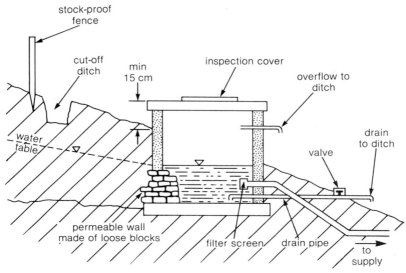

Fig. 9.1 A spring supply should be protected by a catch-pit, which has the features shown in the diagram. The permeable section of the wall should be large enough to allow all the spring discharge to find its way into the chamber without flowing out elsewhere. Water is taken into supply, under gravity flow, through a pipe which is fitted with a filter screen. There is an overflow to control any excess water, and a drain and inspection cover to enable maintenance to be carried out. The inspection cover should be lockable and waterproof, and the overflow pipe should be made vermin-proof to prevent animals or insects from polluting the supply. A cut-off ditch prevents surface water from flowing into the chamber, and animals are kept away by a stockproof fence.

mind of the person who commissioned its construction. These expectations are generally qualified in terms of the quantities of water required. For example, a 'good' borehole for a farm supply may not be capable of supplying a village. The maximum quantity of water which can be pumped from a well or borehole can be reliably assessed only by carrying out a pumping test. It is usual to use a step test, where the pumping rate is increased in up to five roughly equal rates, while the water level in the

pumped borehole is being measured. The pumping rate is plotted against drawdown in water levels to ascertain the maximum rate at which the borehole can be pumped (see Fig. 9.2).

In hand-dug wells, the shallow depth and the relatively small yield often mean that a stepped pumping test is neither practical nor appropriate. The maximum yield can be assessed by a more simple pumping test where the water level is drawn down to just above the pump inlet and then the pumping rate reduced to balance

133

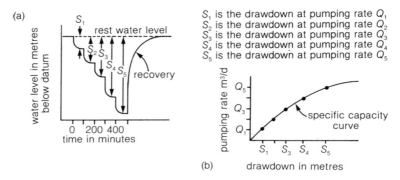

Fig. 9.2 A step-drawdown test should be carried out at roughly equal pumping rates, with the borehole water-level monitored throughout. The drawdown is calculated by subtracting the stable pumping water level achieved at the end of each pumping rate, from the rest water level as shown in (a). A specific capacity curve (b) is constructed by plotting the pumping rate against drawdown. At higher rates of pumping, the amount of drawdown increases, indicating that the maximum yield of the well is being approached.

the inflow to the well. This pumping rate will be the maximum yield of the well, but it may vary throughout the year, however, as seasonal fluctuations in the water table will alter the maximum drawdown values and hence the pumping rate.

9.1.3 Constant-rate tests

Once the maximum rate of pumping has been established, it is prudent to carry out a longer pumping test at a constant rate. Ideally, this should involve continuous pumping for several days, during which time the water levels in the pumped borehole are monitored. This will establish when stable pumping levels have been achieved and will also provide a warning in case the water level should fall too low and damage the pump.

In many instances, it will be pos-

sible to test-pump the borehole much harder than it is likely to be pumped in operational use, because the installed pump capacity will often produce the daily quantity in just a few hours. If the borehole and aquifer stand up to this rigorous testing, the long term reliability can be accepted with greater confidence.

A constant-rate test is also normally required to ascertain whether or not the new borehole abstraction will reduce the yield of nearby wells, boreholes or springs by lowering groundwater levels (see Fig. 9.3). It is possible to calculate the hydraulic properties of the aquifer involved and to calculate these effects, but in many instances the questions are best answered by field measurements.

The duration of the pumping test may vary from three or four days to two or three weeks, depending on the

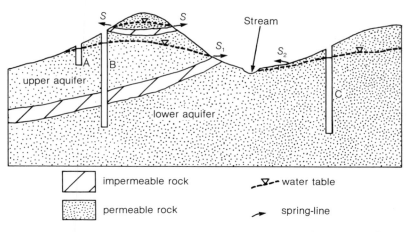

Perched water table

impermeable rock   ▽— water table

permeable rock   ➤ spring-line

Fig. 9.3 Following the construction of a new borehole at *B*, a pumping test is required to ascertain the effects on local sources. A survey reveals spring-lines *S*, S_1 and S_2, a shallow well at *A* and a borehole at *C*. The springs from the perched water table would not be at risk, but careful monitoring may still be needed to demonstrate that any reduced flows are due to the limitations of the aquifer and not to the test pumping. If borehole *B* receives part of its yield from the upper aquifer, then well *A* and spring-line S_1 are at risk. Pumping from the lower aquifer could affect spring-line S_2 and borehole *C*, besides the baseflow of the stream. The extent of these effects would depend on the abstraction rate, the aquifer hydraulic properties and the available groundwater resources.

rate of abstraction. Usually, the higher the required yield, the longer the test. Pumping rates of around 50 m³/d need about 5–7 days test, but 5 000 m³/d or more, is likely to require 14–21 days or even longer. If possible, do not set a precise time limit on the test. Monitor the drawdown in observation boreholes, and continue pumping until water levels have reached an equilibrium. Do not forget that once the pumping-water levels have reached a steady state, fluctuations in water levels caused by factors such as atmospheric pressure

changes will be noticeable again (see Chapter 6); do not mistake them for pumping effects. In some cases it may be appropriate to regard the first year or two of abstraction as a pumping test. Before any pumping test is carried out, identify the main aquifers and locate all the local sources of supply. A monitoring programme should be established which incorporates spring flows and groundwater levels in wells and boreholes. Measurements should start a few weeks before the test and continue for a similar period once pumping has finished. This will enable natural fluctuations and those caused by other existing abstractions to be identified, so that any effects of the new abstraction can be picked up. A detailed description of planning pumping tests is given in *A Field Guide to Water Wells and Boreholes*, Clark (1988).

An important part of investigating a new source is to assess the available recharge by using water balance considerations. Where the new abstraction is small compared with the recharge, the effects are likely to be small, provided that there is sufficient storage within the aquifer. Where storage is small compared with the abstracted volumes, any new pumping is likely to significantly dewater the aquifer, causing problems to existing supplies. All these aspects should be investigated during the pumping-test programme.

9.2   Loss of yield

From time to time a hydrogeologist may be called in to discover why a water supply has failed. The owner usually has an idea of the possible cause, but in many cases his suspicions are based on prejudice and should be treated with both caution and tact. There are many possible causes for a water supply to fail, but in broad terms they can be divided into two; firstly, deterioration of the source itself and secondly, changes in local hydrogeological conditions. The main symptom in all cases is a deterioration in yield, and consequently, it is

very easy to mistake the cause. All water abstraction systems need regular maintenance to prevent the yield from falling off. Unfortunately, this is often ignored and can cause a total failure of the supply. It is human nature to blame others for one's own misfortunes, and often in these cases a nearby borehole abstraction will be held responsible. Pumping from boreholes does lower the groundwater levels, and this may cause springs to dry up, so it is important to ascertain the truth, as compensation may be payable. There are other, less obvious, ways in which groundwater conditions may be changed. These are usually unforeseen consequences of mining, quarrying or civil engineering construction.

9.2.1   Deterioration of spring sources

Spring catch-pits may slowly silt up as find-grained material is washed out of the aquifer formation into the chamber. Some of the fines will settle in the spaces in the catch-pit wall, thereby reducing its permeability. From Darcy's law we know that this will cause the groundwater levels to rise in the immediate vicinity of the catch-pit and, as a result, the groundwater may be able to discharge elsewhere. Look for boggy areas around the catch-pit; they have probably been formed in this way. Cleaning out the catch-pit may be the cure, but where the problem is long standing it may be better to rebuild the catch-pit altogether. In either event, do not forget to advise the owner to disinfect the catch-pit when work has been completed (see section 10.7)

9.2.2 *Deterioration of wells and boreholes*

The yield of wells – but especially boreholes – may slowly decline, caused by clogging of the well face. This can be due to fine-grained particles moving into the well in the same way as we have already considered for spring catch-pits. Clogging can also be caused by the precipitation of minerals on the well face, or by the growth of bacterial slimes. The effect of clogging is to reduce the permeability of the well face, hence, if we again consider Darcy's law, the pumping rate will be reduced for any given value of drawdown. The main symptoms of clogging, therefore, are a decline in yield or an increase in drawdown, or more commonly, a mixture of the two. Although pumping levels have fallen, the rest water levels should recover to values similar to those when the well or borehole was first constructed. These changes in yield characteristics are usually associated with pumping boreholes. The processes which give rise to the problem are not caused by pumping, however, although this may often accelerate it. Do not be surprised, therefore, to discover that the specific capacity of a borehole has declined during a long period of disuse. Chemical precipitation results either from the mixing of two groundwaters with different chemistries, or in response to

changes in groundwater pressure conditions. These processes can occur when a borehole connects different levels within an aquifer that are not naturally interconnected, each having a different pressure and chemistry. Bacterial growth will continue wherever there is a source of nutrition. Iron bacteria are the most common type to affect boreholes and may be provided with adequate food from the steel well-casing, the rising-main and the ferrous components in the pump.

To quantify the deterioration in yield, carry out a step-drawdown test and compare the results with the specific capacity curve produced when the borehole was first drilled, as in Fig. 9.4. In *A Field Guide to Water Wells and Boreholes*, Lewis Clark describes a method of analysing this

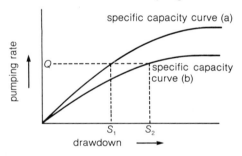

Fig. 9.4 When the well-face of a borehole becomes clogged, the yield deteriorates and the specific-capacity curve (see Fig. 9.2) changes as shown above. When the borehole in this example was first drilled, the yield–drawdown relationship was described by curve (a), but, after clogging, the relationship has changed to that shown by curve (b). The decreased efficiency is shown by considering the change in drawdown at pumping rate Q.

type of pumping test to identify how much of the total drawdown is caused by the inefficiency of the borehole. This is particularly useful when there is little available information from when the borehole was first drilled.

There are a number of cures for clogging problems. They include acid treatment and mechanical swabbing, and it is essential that this work is carried out by an experienced contractor.

9.2.3 Deterioration of the distribution system

Sometimes the failure of a water supply is not caused by a deterioration of the source and it is the pipework and tanks which make up the distribution system that are at fault. Leaking tanks or pipes are a common problem and so is the blocking of pipes with silt or iron oxide (ochre) deposits. It is prudent to include an inspection of the entire system and not just the source, when you investigate a water supply failure. If there are no obvious leaks or blockages, concentrate on the hydrogeological aspects of the job. Should these considerations suggest that there has been no significant change in the source works, and that there are no external causes such as new boreholes in the area, then return to a more detailed examination of the supply system. This may involve careful measurement of water flows through the pipe network and monitoring water levels in covered tanks and reservoirs. For more detailed information on the

maintenance of small water supplies see Brassington (1983).

9.3 Lowered groundwater levels

When groundwater levels are lowered, spring flows will begin to diminish and will eventually cease; the summer flow of streams will reduce; the level of some lakes and ponds may go down and the yield of wells and boreholes will decline. The most common cause of wide-scale lowering of groundwater levels is the abstraction of large quantities of water, although any groundwater pumping may have some of the effects listed above, albeit on a restricted scale. When confronted with solving a problem of reduced yields, etc., examine records of abstractions and groundwater levels over as long a period as possible. Include both rest water levels and pumping-water levels for all the boreholes where records can be found. Figure 6.9 shows an example of how these data can be used. You should also examine records of spring flows and look at the baseflow of streams. Good-quality flow records will be needed for you to separate the baseflow component as described in section 7.7. In this type of study it is important to look at a period of record starting before the suspected abstraction began. Avoid confusing the symptoms of a deterioration of the source works with those of falling groundwater levels, by paying particular attention to long-term trends in groundwater levels changes.

9.3.1 Problems caused by excavations

Pumping for water supplies is not the sole cause of lowered groundwater levels. Quarrying, mining and civil engineering works can all cause significant long-term or even permanent reductions in levels. When a large excavation is made into an aquifer, it increases the void space (i.e. effective porosity, or specific yield) to one hundred per cent. In clean gravel aquifers, this represents an increase of three- or four-fold, but in some aquifers the increase can be as much as a hundred times. Once this large increase in storage has been created below the water table, flow will take place to fill the extra space, thereby temporarily lowering the water table in the vicinity.

The amount by which the water table is lowered depends upon the depth and total volume of the excavation below the water table, the aquifer's hydraulic conductivity, and the specific yield. Groundwater will continue to flow into the excavation until the former water levels are achieved and the original groundwater flow regime is re-established. Where the quarry incorporates an overflow or other form of gravity drainage which dewaters it, water levels will not fully recover. If pumping is needed to dewater the excavation during its working life, it will prolong the period of disruption to

groundwater conditions and will be likely to extend the size of the area affected. Permanent lowering of groundwater levels can occur if the quarrying has the effect of moving the spring line (Fig. 9.5).

It is usual for quarries to be restored once they are complete, either by backfilling to the original ground levels, or by allowing them to flood with groundwater and become a lake. When predictions are being made to determine the level to which a quarry may flood, always use information from the original explorations boreholes, rather than the experience during the excavation period. Where the specific yield is low, it is likely to take a year or two for groundwater levels to build up again.

9.3.2 Deep mining

Almost without exception, deep mines require draining by pumping, because either the mine workings are in water-bearing rocks or the shafts penetrate aquifers before they get down to the working levels. Mining engineers usually attempt to limit the quantity of water entering the mine by sealing off the most permeable rocks. However, large quantities of water are continuously pumped from deep mines, causing dewatering over extensive areas. This has the effect of repeatedly moving the hydrogeological boundaries. Mining also alters the groundwater flow pattern by providing new flow paths via the workings and drainage adits. Subsidence over mine workings may open faults and

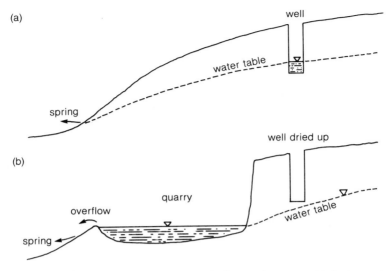

Fig. 9.5 A large excavation into an aquifer lowers the water table by increasing the available storage volume to hold groundwater, and also by dewatering to keep the excavation dry. The effect will be permanent unless the quarry is allowed to flood to the original water-table level. The top diagram (a) shows groundwater conditions prior to quarrying. After the quarry in this example was excavated it was allowed to flood (b), but groundwater levels are controlled by the overflow and kept at a lower level than before. This has caused the well to dry out, but will have little effect on the spring flows. There may be a deterioration in spring-water quality, however, if the water in the quarry becomes polluted.

fissures which will increase permeability on a local scale.

Hydrogeologists may be involved in groundwater problems with deep mines from two points of view. Investigations to demonstrate that water supply losses have been caused by mine dewatering, should be investigated in a similar way to supply problems. To predict how much water will have to be pumped from a proposed new mine, you will need to drill test boreholes, carry out pumping tests and probably use computer-

modelling techniques, all of which are beyond the scope of this Handbook.

9.3.3 Civil engineering works

Many civil engineering construction projects involve short-term excavations, some of which are below the water table. These can have the same effects as outlined in section 9.3.1, but usually construction takes place over a relatively short period, which reduces the impact on groundwater

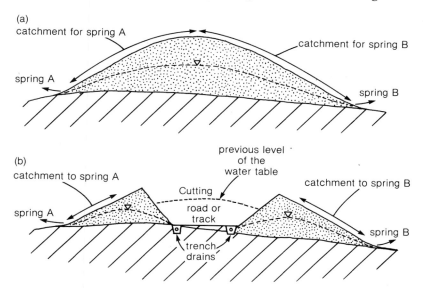

Fig. 9.6 A railway or road-cutting has been excavated through a hill which contains an aquifer from which groundwater issued at springs *A* and *B*, (a). Once the cutting was excavated below the water table, it was kept dry by trench drains installed on both sides, (b). This significantly reduces the flow of both springs by reducing their catchment areas by more than 50 per cent. (Note: trench drains are frequently installed at the top of cuttings to help maintain slope stability, but these have been omitted from the diagram.)

supplies. In some types of construction, permanent groundwater drainage systems are incorporated into the works, either to prevent groundwater pressures building up which may alter the stability of slopes, the bearing properties of rocks and soils, or to prevent the flooding of cuttings and tunnels.

When deep cuttings are made during the construction of a new railway or road, it is standard practice for large-capacity drains to be incorporated. These drains usually run along the bottom of each side of the cutting and discharge into local streams where the topography allows. The drains allow groundwater to flow along them, thereby causing permanent depletion of groundwater levels (see Fig. 9.6). The construction of tunnels or even major pipelines can have the same result, by providing a means whereby groundwater can flow along the outside of the tunnel lining or via the pipe bedding material. This can result in some aquifers being drained, with the consequential disruption to water supplies. To minimize the disruption, all local supplies should be located before work on the tunnel or pipeline project starts. A

141

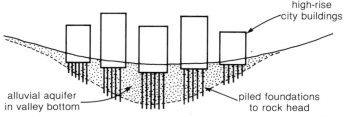

Fig. 9.7 The piled foundations of high-rise buildings in urban areas may be sufficiently dense to reduce the effective cross-sectional area of a shallow aquifer. This will cause groundwater levels to rise on the 'upstream' side of the buildings, and could affect the yield of nearby wells.

hydrogeological assessment should then be made to identify those sources at risk, followed by a monitoring programme which should be started before the work begins; this means that any changes caused by the project can be clearly seen. Groundwater flow along backfilled pipeline trenches can be reduced by constructing antiseepage collars around the pipe. These are built by excavating a series of short cuts across the main trench and at right angles to it. The cuts should be at least five pipe diameters long and at a similar depth. They are backfilled with mass concrete to near ground level. Construct a series of half a dozen or so, spaced at intervals equal to ten pipe diameters along critical lengths of the pipe.

Piled foundations, especially in urban areas, can radically reduce the cross-sectional area of shallow aquifers and may result in groundwater level changes (see Fig. 9.7).

9.3.4 Land drainage

In shallow aquifers, the deepening of ditches or streams and the installation of a field drainage system will usually lower the water table in the immediate area. Figure 9.8 shows how this can happen. In most cases the cure is to deepen the affected wells or replace spring supplies with a well or borehole.

9.4 Rising water tables

Up to now, we have been considering changes brought about by falling groundwater levels. In recent years, however, there have been an increasing number of incidents where water tables have steadily risen, causing flooding problems to the basements of buildings and other deep structures such as railway tunnels, or even flooding of low lying land. In most cases the cause is a significant reduction in groundwater pumping, and this can be established by studying the relevant records. Rising water tables are often found in older urban areas, where ageing water mains and sewers may leak, adding to the problem.

(a)

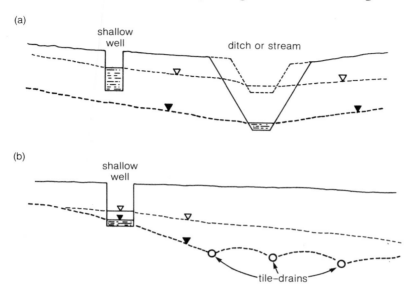

(b)

Fig. 9.8 The improvement of land drainage can cause permanent lowering of a water table, with consequential reduction in well yields. In example (a), a drainage ditch (or stream) has been regraded and deepened. The ditch acts as a discharge point for groundwater in the area, and therefore deepening it has lowered the water table, which, in turn, has dried out a shallow well. In example (b), the installation of a tile-drain system has had the effect which the farmer wanted, and has improved the field drainage by lowering the water table. This has dropped the water level in a nearby shallow well, thereby reducing the yield. In both cases the remedy would be to deepen the affected wells.

Leaking sewers may also cause pollution problems, which in turn could lead to a reduced rate of abstraction.

There have been a limited number of cases reported, where the rising water table has been caused by excessive irrigation of crops and even parks and gardens. To decide if this could be the cause of a rising water table, examine irrigation records and compare the rate of application with the likely transpiration rates.

If you are asked to find the reason why a basement is flooding, do not jump to the conclusion that it is caused by a rising water table, even in areas where groundwater level changes have occurred. The most likely causes are broken water mains or sewer pipes, and this can be established from a chemical analysis of the floodwater. Check the analysis against those for local mains water and from groundwater samples taken from a nearby well. If the problem is caused of a leaking sewer, the water is likely to have a high nitrogen content in the form of ammonia, nitrite or nitrate, depending on the oxygen content.

9.5 Waste disposal sites

The most widely employed method for the disposal of domestic, commercial or industrial waste is to tip it into excavations or on to land, a technique often referred to as *landfill*. Most waste materials pose a threat to water quality, because rain-water percolating through the waste produces a noxious liquid termed *leachate*. The pollution potential of leachate varies from one waste material to another, and also depends upon the hydrogeological conditions of the landfill site.

Mining wastes may produce a leachate with high concentrations of minerals associated with the mining activities. Other wastes, which include domestic refuse, trade and industrial wastes, often produce a high-strength organic leachate which has the potential for causing long-term groundwater pollution. Daly and Wright (1982), Freeze and Cherry (1979) and Naylor *et al.* (1978) describe how such leachates are formed, and provide detailed information on the likely composition of a variety of leachates. Table 9.2 provides an indication of the ranges in concentrations of the major constituents found in domestic waste leachates. In general, the lower concentrations refer to older wastes, as leachate strengths decrease with time. As leachates break down, methane and other gases may

be produced. They can present a hazard of explosion but, where properly managed, these gases can be collected and used as an energy source.

An investigation of the geological and hydrogeological conditions of all proposed landfill sites should be carried out before a decision is taken on the suitability of the site. The study should include information on permeabilities; groundwater levels; fluctuations and flow directions; the thickness of the unsaturated zone; and the mineralogy of the geological materials on the site. This will enable the probable rates of leachate movement to be calculated and a judgement to be made on whether there is likely to be an attenuation of the leachate as it flows through the underlying rocks. The investigation should assess the availability of materials on-site for use in any preparatory engineering works, and also consider the risk of methane gas migration away from the site.

The investigation should follow the various stages set out in this Handbook, and is likely to include a drilling phase. Give careful thought to both the number and location of investigation boreholes, and also to the drilling methods. Ensure that all boreholes which are not required for long-term monitoring are properly backfilled with cement or bentonite grout, to prevent leachate migration along them. It is a good practice to avoid drilling the boreholes underneath the area where tipping is to be done.

Before you start drilling boreholes,

it is a good idea to contact the local pollution control authority and agree with them the details of the investigation.

Table 9.2 Typical concentrations of the major chemical components found in leachate from domestic waste

Parameter	Typical range of values
pH value	6.0–8.0
Biological oxygen demand (BOD)	100–12 000
Chemical oxygen demand (COD)	500–60 000
Total organic carbon (TOC)	100–20 000
Ammonia (as N)	10–1000
Nitrite (as N)	0.2–2.0
Nitrate (as N)	0.5–5.0
Chloride (as Cl)	100–3000
Sulphate (as SO_4)	50–1000
Sodium (Na)	40–2000
Potassium (K)	20–2000
Magnesium (Mg)	10–500
Calcium (Ca)	1.0–180
Manganese (Mn)	0.5–200
Iron (Fe)	0.1–2000
Copper (Cu)	0.01–0.3
Nickel (Ni)	0.05–1.5
Lead (Pb)	0.05–8.0
Zinc (Zn)	0.05–130

All values in mg/l except pH; based on data from various sources.

10
Good working practice

Many aspects of hydrogeological fieldwork may bring a hydrogeologist into contact with potential hazards. It is very important to appreciate the inherent dangers so that you avoid taking unnecessary risks. This brief note can only alert you in a general way as to what constitutes a potential hazard. It is up to *you* to ensure that you have the proper equipment and training and that you follow appropriate safe practices in carrying out your field measurements. Remember that *hazards* cover a wide range of dangers which could affect you, your colleagues helping you in the field, or members of the public.

10.1   Safety codes

In Britain, field safety is covered by the Health and Safety at Work Act, 1974, which gives obligations to both employers and workers to ensure that safe working methods are adopted. The various government and local government agencies involved in hydrological fieldwork usually have their own safety codes. Whichever country you are working in, make

sure that you read and understand any safety codes that your employer has issued. If an appropriate code is not available in this way, borrow one from the local water authority or similar organization and use that.

There are many hazards involved in working in remote locations, particularly on your own, when even a minor accident or incapacity may become serious if loss of mobility prevents aid from being summoned. These risks are similar to those encountered by hill walkers and others who work in remote areas. There are a number of good books which contain sound advice on safety precautions emergency procedures for these remote areas. A selection of these books is given in the reference list.

Before you set out to work in a remote location, leave details of your route and work location with an appropriate person at your base. Include approximate times of your return and details of what action should be taken if you fail to turn up. Do not forget to check in – whatever time you get back – to prevent abortive search and rescue activities.

Before setting out, check the weather forecast on the television, radio or telephone information service, and avoid working in severe weather conditions. There is little point in going out and 'braving' the weather. Bad weather is likely to reduce the amount of work that you can complete and, by putting yourself at risk, you are placing an unnecessary strain on the rescue services.

A great deal of work is likely to be carried out in rural areas, so always adhere to the code set out in Table 10.1. Never enter private land without first seeking permission from the landowner. Always be on the look-out for livestock, particularly bulls, which could be an additional hazard to your fieldwork.

Table 10.1   Code for minimum impact of fieldwork on the countryside

---

1   Risk of fire
Woodlands, grasslands, heaths, etc. can be destroyed by fire. Take great care to control any fires you may need for cooking or warmth. Always ensure that they are properly out before leaving them. If you smoke, take care not to drop burning matches, cigarette ends or pipe ash.

2   Livestock
Close all gates, even if you find them open, to prevent livestock from wandering. Do not force your way through fences, walls or hedges, as any damage could allow animals to escape. Do not chase animals; they may injure themselves, or could turn on you and cause you injury. If you take a dog with you, prevent it from chasing or frightening animals. Loud noise, the playing of radios or cassette players will also disturb animals.

3   Crops
Do not walk through crops – even grass – unless you have express permission of the owner. Stick to recognized footpaths or the edges of fields.

4   Litter
Do not drop any litter of any sort. Take it home with you, or back to your base. When out for several days, acceptable alternatives are burning (watch for fires) or burial.

5   Protect water supplies
Do not let any material enter a water supply. This is especially important when taking measurements or samples.

6   Wild life
Do not disturb wild animals or birds. Do not pick wild flowers or damage trees or other plants. Some wild animals can be dangerous, so find out about

any which may be in your area. Do not eat wild plants or berries unless you know that they are not poisonous.

7 Country roads

Country roads have special dangers, such as blind corners, high banks, animals, farm vehicles or machinery and often poor surfaces. Take care when driving and if walking, face oncoming traffic.

8 General

Set a good example and try to fit in with the life and work in the countryside. This effort is especially important when you are working in a foreign country. Not only is this behaviour basically good manners, but it will be invaluable in developing good relationships with the landowners in your study area.

10.2 Clothing and equipment

Always ensure that your footwear and clothing is suited to the weather, the terrain and to the task that you are undertaking. The basic principle is to remain at a comfortable body temperature, that is, warm and dry in cold areas and to avoid overheating in tropical countries. If necessary, carry a spare sweater and waterproof clothing, or a sun-hat, in a rucksack. For much of the time, good quality walking boots are the best form of footwear. Boots give a good grip when walking over rough ground, and provide protection to ankles – important where there may be poisonous snakes, scorpions or spiders. On a drilling site, wear safety-boots or shoes, and when you are carrying out measurements in a stream, thigh-waders will be required. Ensure that the soles of your waders have a good grip, and do not be tempted to keep old ones with a smooth sole, even though they do not leak. Other items of personal safety equipment include helmets, which should always be worn in the vicinity of rock faces and on drilling sites; gloves, goggles and ear protectors for when you are near noisy machines.

When working in remote places, your personal emergency equipment should include a first-aid kit, a whistle or a flashlight for signalling. Do not forget to take spare batteries, matches sealed in a waterproof plastic bag, and a survival bag. This is a large, brightly coloured, heavy-duty polythene bag available from all outdoor sports shops. If you have to spend the night out unexpectedly, simply climb into the survival bag, which will keep you dry and help keep you warm. However, the bag can make you sweat; this is dangerous as the sweat can freeze. You should not sit on the cold ground because the survival bag will not provide any insulation. Sit on your rucksack or a heap of leaves and branches to keep you off the ground. You should also carry some form of emergency rations. These could be the bitter 'sportsman's' chocolate which will deter you from eating it until really necessary, glucose tablets, or one of the many dried-fruit mixtures recommended in mountain safety books. A water-bottle or effervescent water-purifying tablets should also be carried. Do not overburden yourself unnecessarily by carrying a large rucksack of equipment everywhere. Give careful consideration to the safety aspects of each field-trip and take the appropriate gear with you.

10.3 Distress signals

The accepted emergency distress signal is six blasts on a whistle or six flashes with a mirror or torch, repeated at one minute intervals. Rescuers reply with only three blasts or flashes, again repeated at one minute intervals, which are intended to prevent rescue parties from homing in on each other. If you are in the unfortunate position of needing to make these signals, remain calm and wait the full minute between signals. Panic may cause signals to be sent at intervals as short as ten or fifteen seconds, with potential confusion resulting. Visual signals can be improvised with an inflated survival bag which can be seen by parties, or large letters in the

form of SOS which will be visible from the air. Be careful if you decide to use a signal-fire – keep the fire under control.

10.4 Exposure

All people working in temperate or cold climates in mountainous areas, particularly when they may get wet, should learn the dangers of exposure. This is the common name for *hypothermia*, which is deep-body cooling and is not uncommon in hilly areas in almost any part of the world. It is not confined to mountains, and it is not limited to the winter months; sudden drops in temperature can occur at any time of the year on any high or exposed ground. Getting wet, either by being caught out in a storm or by falling in a stream will reduce body temperatures and can bring on hypothermia. It is important to recognize the symptoms, both in yourself and in your companions, and to know how to treat it. The signs and symptoms in the early stages are white or pale complexion, violent shivering and complaints of feeling cold. Characteristically, the victim loses interest in what is happening, slows down and may even stop complaining of the cold. This means that if you are on your own, you are likely to be unaware of the onset of hypothermia unless you consciously keep it in your mind *as soon as you start to feel cold*. As the blood going to the brain cools further, judgement will be impaired, abnormal or irrational behaviour may be displayed – slurring of speech, disturbed vision, stumbling and falling may also occur. These are all very serious symptoms; as soon as they are recognized take shelter immediately and get the victim out of the wind. Try to get dry clothes on them if possible, but at least cover them with windproof materials such as the survival bag in your emergency kit. If possible give them a hot sugary or glucose drink but – *do not give them alcohol – this could kill them.* Alcohol dilates the smaller blood vessels so that blood flows to the extremities more rapidly, which will accelerate heat loss. Never underestimate the seriousness of hypothermia. Death can occur within one hour of the onset of symptoms. People's tolerance of cold varies enormously, so take no chances. Whenever possible seek immediate help. Look out for cases of incipient exposure in other members of your party.

To keep the victim warm, it may be necessary to get into the same sleeping bag and use your body heat. If the victim is in a state of total collapse, get help quickly, for if their temperature drops below 31°C (88°F), only medical treatment can save them. If you carry them on a stretcher, keep their head lower than their feet. Back at base, put them fully dressed (to reduce shock) into a bath at 45°C (113°F) for 20 minutes, *providing* that their temperature is not below 31°C. If they are colder than this then get them to hospital, but, if all else fails, *allow them to warm up slowly* in a warm room.

10.5 Heat exhaustion

Heat exhaustion is due mostly to dehydration (water loss) and leads to a shocked condition, giving rise to rapid pulse, cold clammy skin, thirst, fatigue and giddiness. If untreated it can lead to delirium and coma. Urine output is low. If similar symptoms exist, coupled with muscle cramps and no marked rise in body temperature, then salt deficiency is also a factor. Treat the patient by placing him in a cool, shady place. Restore the water and salt balance by giving the patient lots of cool water, containing half a teaspoon of salt per pint, to drink.

Heat-stroke is the failure of the body's temperature regulating mechanism. The onset is sudden, and may be preceded by heat exhaustion. It is very serious, as the body temperature is likely to be increasing all the time. The main symptoms are high body temperature, hot dry skin, the *absence* of sweating, possibly aggressive behaviour and lack of co-ordination. Convulsions, coma and death are likely to follow unless effective treatment is given. Obviously, the patient must be cooled down by whatever means are available – for example, removal of clothing and splashing tepid water on to the body, or generating air movement by fanning. If water is scarce, lay wet clothes on the patient and create air movement over him to lose heat by evaporation of the water. Once the patient has started to recover, make him rest and give cool salt drinks as described above.

Avoid heat-stroke by not working during the hottest part of the day; sit in the shade instead. Do not remove your clothes, as keeping them on prevents the sweat evaporating too fast. Wear light coloured clothes because they reflect the heat. Eat little and often, as the digestion of large meals uses up a lot of water. Drink often, but boil water or use purification tablets, and do not drink sea-water.

10.6   Working near wells and boreholes

There are two main hazards to bear in mind when working around boreholes and wells. These are the dangers of falling into the well or large diameter borehole, and problems caused by the generation of gases. There are other hazards to remember, such as the trapping of hands and feet and the danger of back-strain when removing heavy covers. Underground chambers and the dark corners around well-heads may be an attractive home to snakes or poisonous insects. Bear this in mind before putting your hand in.

When working on boreholes, wells, chambers or shafts where the access to the hole exceeds 15 inches (375 mm), either reduce the size of the opening to prevent any risk of slipping and falling through the opening, or wear a safety-harness with a lifeline attached to a secure point. In most cases, lifelines should be short enough to prevent a free fall of more than three metres, and be either secured to an immovable

object or held by at least two 'top men'. If you are going to use a ladder or step-irons, make sure that they are secure before you trust your weight on them.

Before you carry out any work over chambers or wells, carry out a visual examination of cover plates, staging, etc. Beware of corroded metalwork or rotten and/or slippery timber. Remember that apparently sound timber can be rotten underneath. If in doubt, do not proceed until appropriate remedial work has been carried out, or use a safety-harness and lifeline.

Various gases may emanate from an open borehole. These include carbon dioxide, which can cause suffocation; hydrogen sulphide, which is highly toxic and methane, which has inherent dangers of explosion. If the top of the borehole is above ground level and is in the open, then the dangers posed by these gases are very low indeed. However, where the borehole is in a chamber or in the basement of a building, then great care must be taken. Use an appropriate gas detector or miner's safety lamp to test the amount of oxygen in any confined space before you enter it. You should also test the atmosphere with a methane meter or gas alarm, to see if methane is present in concentrations which could cause an explosion. Barometric conditions are important as atmospheric pressure strongly influences the emission of gases. It is usually safe to enter a chamber if the barometric pressure is rising, as air will be entering the chamber, but do

151

not omit other tests. It is usually dangerous to enter a chamber when the pressure is falling, as gases will be flowing out of the borehole into the chamber. If in doubt, do not enter the confined space on your own or without proper safety equipment.

Methane is also produced by the decomposition of refuse in waste disposal sites. If you are carrying out a sampling exercise in or around a waste-disposal site, be particularly vigilant for this gas.

10.7 Hygiene precautions

Please remember to be very careful when taking any readings on a well, borehole or spring which is used as a drinking water supply. Avoid knocking dirt into the well or catch-pit when removing the cover, and ensure that all the equipment that you use is clean. If in any doubt, disinfect it all by a thorough washing in a chlorine solution – made up by mixing a litre of proprietary chlorine based disinfectant which contains between 10 and 20 mg/litre of chlorine with five litres of water. Suitable brands include Chloros, Domestos and Parazone, which are all made with sodium hypochlorite (NaOCl). Do not use 'pine' disinfectants. These are based on phenols, which are easily detected by taste, and only very small quantities (1 mg/l), will contaminate a water source.

Wash the buckets and jugs used in spring-flow measurement with the chlorine disinfectant, both inside and

out, before putting them in a catch-pit. Clean the probe and tape of any dippers which are to be used on supply wells or boreholes. Samplers should also be cleaned in the same way, and do not forget the sampler cable. If it is necessary to put your hands into drinking water, try to wear rubber gloves, or at least wash your hands thoroughly.

If you are measuring water levels or taking samples around a waste-disposal site, or anywhere else where groundwater may be contaminated, reserve certain equipment for that work. *Never use these instruments on drinking water supplies.*

Newly constructed catch-pits and wells are often disinfected before being put into use – with a chlorine-based solution. Fill the well or catch-pit with the disinfectant, and allow it to stand overnight. Then pump all the water out, and continue pumping until any chlorine smell has disappeared.

10.8 Trial pits

When taking samples or measurements from a trial pit, never climb into an unsupported excavation which is greater than 1.25 metres deep. Even shallow trial pits can suddenly collapse without warning, especially if groundwater has been encountered. Care must be exercised when approaching an excavation and such approaches kept to a minimum. Once logging has been completed, the excavation should be backfilled as soon as possible, or fenced off and clearly marked if left unattended.

10.9   Waste disposal sites

When working on or around a waste-disposal site there are precautions additional to those mentioned elsewhere in this chapter. It is important not to breathe in harmful vapours or gases and to avoid physical contact with noxious chemicals, which may be on the site either as solids or dissolved in groundwater. It is important to wear protective overalls, rubber boots and rubber gloves when handling water samples taken from on or near a waste-disposal site. Be particularly careful in sampling leachate, to avoid splashing it on your skin – you do not know what it contains before it is analysed, and some chemicals can be absorbed into the body through the skin. Take care when transporting such samples, so that you avoid coming into contact with them in the event of a car-crash or other accident.

It is suggested that anti-tetanus and typhoid injections should be given, as a precaution to all personnel working in the vicinity of waste disposal sites. Where the site receives sewage sludge, there is the additional hazard of contracting leptospiral jaundice (Weil's Disease). Avoid direct contact with the sludge and report the possibility of this disease to your medical practitioner in the event of *any* illness, as the symptoms can easily be confused with more common infections.

10.10   Current meter gauging

The major hazard that distinguishes this particular type of activity is the absolute proximity of water and the attendant risk of drowning. Do not attempt a current meter gauging by wading, if the velocity of the water or the unevenness of the bed would make your balance precarious. It is best to gauge in teams of at least two people, with one person remaining on the bank with an emergency lifeline ready at all times. Safety harnesses and lines should not be worn when gauging. If conditions are such that a line is thought necessary, then gauging by wading should not be attempted.

Appendix I
Units

It is becoming common for hydrogeological field measurements to be made using metric units, although in some countries the British (imperial) or American systems are still in use. The metric units used do not always conform to the SI system (Système Internationale d'Unités), which is based on the metre (m), kilogram (kg) and second (s). The following comments and conversion tables have been included to help to avoid confusion.

AI.1   Volume

The SI unit of volume is the cubic metre, but the litre (10^{-3} m^3) is an accepted alternative. The megalitre (Ml) which is 10^6 litres or 10^3 m^3, is sometimes used where large volumes are involved. The imperial and American systems both use gallons, but be careful as they are not the same volume; the US gallon is only 0.8326 imperial gallons. If you are using records in gallons, check carefully to identify which gallon is being used.

AI.2   Aquifer characteristics

Intrinsic permeability is usually expressed in darcys (see section 5.1). 1 darcy = 0.835 m/d and 1 m/d = 1.198 darcy for water at 20°C. It is usual for hydrogeologists to express hydraulic conductivity in metres per day (m/d), and transmissivity in square metres per day (m^2/d). This use of the day rather than the second produces values of the aquifer in whole numbers. The imperial and American units for hydraulic conductivity are gallons per day per square foot, and for transmissivity, gallons per day per foot. Beware of the different sized gallons.

AI.3   Pumping rate

The SI unit of discharge is the cumec (m^3/s), but this unit is only convenient for

154

large surface flows. Litres per second (l/s) or m³/d are more practical units for the flow of small streams, springs and borehole discharges. The latter unit also conforms with the m/d and m²/d unit of hydraulic and transmissivity.

AI.4   Conversion factors

AI.4.1   Length (SI unit, metre, m)

	m	ft	in
1 m	1.000	3.281	39.37
1 ft	0.3048	1.000	12.00
1 in	2.540×10^{-2}	8.333×10^{-2}	1.000

AI.4.2   Area (SI unit, square metre, m²)

	m²	ft²	acre	hectare
1 m²	1.000	10.76	2.471×10^{-4}	1.0×10^{-4}
1 ft²	9.29×10^{-2}	1.000	2.29×10^{-5}	9.29×10^{-6}
1 acre	4.047×10^{2}	4.356×10^{4}	1.000	4.047×10^{-1}
1 hectare	1.0×10^{4}	1.076×10^{5}	2.471	1.000

AI.4.3   Volume (SI unit, cubic metre, m³)

	m³	l	imp. gal	US gal	ft³
1 m³	1.000	1.000×10^{3}	2.200×10^{2}	2.642×10^{2}	35.32
1 l	1.000×10^{-3}	1.000	0.2200	0.2642	3.532×10^{-2}
1 imp. gal	4.546×10^{-3}	4.546	1.000	1.200	0.1605
1 US gal	3.785×10^{-3}	3.785	0.8326	1.000	0.1337
1 ft³	2.827×10^{-2}	28.27	6.229	7.480	1.000

AI.4.4   Time (SI unit, second, s)

	s	min	h	d
1 s	1.000	1.667×10^{-2}	2.777×10^{-4}	1.57×10^{-5}
1 min	60.00	1.000	1.667×10^{-2}	6.944×10^{-4}
1 h	3.600×10^{3}	60.00	1.000	4.167×10^{-2}
1 d	8.640×10^{4}	1.440×10^{3}	24.00	1.000

AI.4.5   Discharge rate (SI unit, cubic metre per second, m^3/s)

	m^3/s	m^3/d	l/s	imp. gal/d	US gal/d	ft^3/s
1 m^3/s	1.000	8.640×10^{4}	1.000×10^{3}	1.901×10^{7}	2.282×10^{7}	35.313
1 m^3/d	1.157×10^{-5}	1.000	1.157×10^{-2}	2.200×10^{2}	2.642×10^{2}	4.087×10^{-4}
1 l/s	1.000×10^{-1}	86.40	1.000	1.901×10^{4}	2.282×10^{4}	3.531×10^{-2}
1 imp. gal/d	5.262×10^{-8}	4.546×10^{-3}	5.262×10^{-5}	1.000	1.201	1.858×10^{-6}
1 US gal/d	4.381×10^{-8}	3.785×10^{-3}	4.381×10^{-5}	0.8327	1.000	1.547×10^{-6}
1 ft^3/s	2.832×10^{-2}	2.447×10^{3}	2.832×10^{4}	5.382×10^{5}	6.464×10^{5}	1.000

AI.4.6   Hydraulic conductivity (SI unit, cubic metre per second per square metre, m/s)

	m/s	m/d	imp. gal/d-ft^2	US gal/d-ft^2	ft/s
1 m/s	1.000	8.640×10^{4}	1.766×10^{6}	2.121×10^{6}	3.80×10^{2}
1 m/d	1.157×10^{-5}	1.000	20.44	24.54	4.398×10^{-3}
1 imp. gal/d-ft^2	5.663×10^{-7}	4.893×10^{-2}	1.000	1.201	1.858×10^{-6}
1 US gal/d-ft^2	4.716×10^{-7}	4.075×10^{-2}	0.8327	1.000	1.547×10^{-6}
1 ft/s	2.632×10^{-3}	2.273×10^{2}	5.382×10^{5}	6.463×10^{5}	1.000

AI.4.7   Transmissivity (SI unit, cubic metre per second per metre, m^2/s)

	m^2/s	m^2/d	imp. gal/d-ft	US gal/d-ft	ft^2/s
1 m^2/s	1.000	8.640×10^{4}	5.793×10^{6}	6.957×10^{6}	1.161×10^{2}
1 m^2/d	1.157×10^{-5}	1.000	67.05	80.52	1.343×10^{-3}
1 imp gal/d-ft	1.726×10^{-7}	1.491×10^{-2}	1.000	1.201	1.858×10^{-6}
1 US gal/d-ft	1.437×10^{-7}	1.242×10^{-2}	0.8326	1.000	1.547×10^{-6}
1 ft^2/s	8.617×10^{-3}	7.445×10^{2}	5.382×10^{5}	6.463×10^{5}	1.000

Appendix II
Useful addresses

Chapter 3 describes the wide range of existing information required for a hydrogeological investigation. Most of these data will have been collected by government departments, local government organizations and a variety of other bodies – for a large number of different purposes, most of which have little to do with groundwater studies. This makes your task of obtaining the data you need more difficult, but this Appendix is intended to help by listing the type of organizations which may hold useful information, and by presenting some addresses. This list is by no means exhaustive and, inevitably, will become out of date. Please treat it as a general guide and expect to spend some time in locating the records.

Access to information varies tremendously from one country to another, and a lot will depend upon the status of your employer. If you are working for a government agency it is likely that you will obtain all the maps and records you need. The other extreme is to be met with a firm refusal or such a complex bureaucratic system that you will never get through it. Use your diplomatic skills to the full, be patient and observe local customs. Remember, your views and attitudes may not always be shared by people in other countries. Sometimes politics may cause complications as, for example, where neighbouring countries dispute their common boundary and both lay claim to the same area of land.

If you are unable to obtain information from organizations within a particular country, you may be able to obtain it from United States or European organizations. The geological surveys and university departments from these countries have undertaken overseas work for many years. Besides providing maps and reports, these organizations may help you to locate a source of information in the country where you are working. For example, the USGS publish a 'Worldwide directory of national Earth science agencies and related international organizations' as USGS Circular 834.

There are a large number of different organizations which you may need to approach. The following list is intended to provide an aide-mémoire.

Government Departments	Military, Air Force, Environment, Natural Resources, Mines and Mineral Resources, Agriculture, Water, Forestry, Geological Surveys, Land Surveys, Research Organizations
Local Government	County and District Councils, Local Authorities, Water Boards, City Councils, River Authorities and Catchment Boards, Highway and Roads Authorities, public records office, libraries
Other	Major water-users, industrial and trade associations, consultant engineers and geologists, geological societies, natural history and philosophical societies
Universities and Colleges	Departments of Geology, Geography, Environmental Science, Chemistry, Physics, Mining, Civil-Engineering, Surveying

Useful addresses:

Great Britain

Ordnance Survey
Romsey Road
Maybush
Southampton SO9 4DH

Topographic maps, geological maps

British Geological Survey
Keyworth
Nottingham NG12 5GG

Geological maps, reports and general data for Great Britain and overseas

British Geological Survey
Hydrogeology Unit
Maclean Building
Crowmarsh Gifford
Wallingford
Oxfordshire OX10 0RA

Hydrogeological maps and reports for Great Britain and overseas

158

For Regional Water Authorities in England and Wales and River Purification
Boards in Scotland see local telephone directories or *The Geologist's Directory*.

Ireland

Ordnance Survey Office
Phoenix Park
Dublin

Topographic maps, remote-sensing

Geological Survey of Ireland
Beggars Bush
Haddington Road
Dublin 4

Geological maps and records,
hydrogeological information, and
remote-sensing

An Foras Forbatha
(National Institute for Physical
Planning and Construction
Research)
St Martin's House
Waterloo Road
Dublin 4

Hydrometric records and abstraction
information

Federal Republic of Germany

Kuratorium für Wasserwirtschaft
Kennedy Allee 62–70
D-5300 Bonn 2

Water resources records, etc.

France

Ministère de l'Environnement
14 Boulevard du Général Leclerc
9252 Neuilly-sur-Seine

Water resources records, etc.

Spot Image
16 Bis Avenue
Edourd Belin BP 4359
31030 Toulouse
Cedex

Satellite imagery

Field Hydrogeology

Italy

Ministry of Public Works   Water resources records, etc.
Piccia Porta Pia 1
00100 Rome

United States of America

National Cartographic   Free information on national and state
  Information Center (NCIC)   topographic maps, geological maps and
US Geological Survey   information and remote sensing
507 National Center
Reston
Virginia 22092

Operations Section WRD   Data on surface water, groundwater and
US Geographical Survey   water quality collected by the US
405 National Center   Geological Survey
Reston
Virginia 22092

National Water Data Exchange   National confederation of water-
US Geological Survey   orientated organizations data-share
421 National Center   system
Reston
Virginia 22092

United States Geological Survey   USGS maps, books, professional papers
Box 25425, Federal Center   and other publications on geology and
Denver   hydrogeology of the USA and overseas
Colorado 80225

Information is available from state geological surveys, departments of mines
and departments of natural resources. See NCIC for addresses.

Canada

Energy, Mines and Resources   Topographic maps and remote sensing
  Canada
580 Booth Street
Ottawa
Ontario K1A 0E4

Geological Survey of Canada 601 Booth Street Ottawa Ontario K1A 0E8	Geological maps and records

Information is also available from provincial government departments. Obtain addresses from one of the above or by contacting the government offices in each provincial capital.

Australia

Division of National Mapping PO Box 31 Belconnen ACT 2616	Topographic maps and remote sensing
Bureau of Mineral Resources PO Box 378 Canberra ACT 2601	Geographical maps and records, some groundwater data

For additional information contact the Surveyor General, Department of Lands, or the State Geological Survey in each state or territorial capital, including Tasmania and Papua.

New Zealand

Map Centre Department of Lands and Survey PO Box 6452 Te Aro Wellington	Topographic maps and remote sensing
Geological Survey Department of Scientific and Industrial Research Private Bag Wellington	Geological maps and records

National Water and Soil
  Conservation Authority
Ministry of Works and
  Development
PO Box 12–041
	Publications on water resources topics

161

Catchment Authorities Association PO Box 5054 Lambton Quay Wellington	Central organization for the catchment authorities

Information on groundwater and water resources can be obtained from the individual catchment authorities in the CAA.

India

Survey of India Dehra Dun Uttar Pradesh	Topographic maps and remote sensing
Geological Survey of India Pushba Bhawan Madandir Delhi 62	Geological maps and reports

Sri Lanka

Survey Department PO Box 506 Colombo	Topographic maps and remote sensing (restricted access)
Geological Survey Department 48 Sir Jinarathana Mawatha Colombo 2	Geological maps and records
Water Resources Board 2A Gregory Avenue Colombo 7	Some hydrometric measurements and abstraction records

Bangladesh

Survey of Bangladesh Tejgaon Industrial Area Dhaka	Topographic maps and remote sensing (restricted access)

Bangladesh Water Development
  Board
Rivers Section
Green Road
Dhaka

Limited geological information and
hydrometric records

Bangladesh Agricultural
  Development Corporation
Survey and Investigation Division
Dilkusha C/A
Dhaka

Limited geological records and
hydrometric records

Saudi Arabia

Air Survey Department
Ministry of Defence
Riyahd

Topographic maps and remote sensing

Ministry of Petroleum and
  Mineral Resources
Jeddah

Geological maps and records

Ministry of Agriculture and Water
Water Resources Development
  Department
Riyahd

Hydrometric information abstraction
records

Bahrain

Ministry of Commerce and
  Agriculture
Manama

Sultanate of Oman

Ministry of Agriculture, Fisheries,
  Petroleum and Minerals
Water Resources Department
PO Box 5036 (RUWI)
Oman

Field Hydrogeology

Arab Republic of Yemen

Ministry of Agriculture
Sana'a

Kuwait

Ministry of Electricity and Water
Water and Gas Department
Groundwater Section
PO Box 5395
Kuwait

People's Democratic Republic of Yemen

Ministry of Agriculture and
  Agrarian Reforms
Department of Irrigation
PO Box 1161
K/sr–Aden

Qatar

Ministry of Electricity and Water
Water Department
Groundwater Resources Section
PO Box 162
Doha

Republic of Iraq

Ministry of Agriculture and
  Agrarian Reforms
Baghdad

United Arab Emirates

Ministry of Agriculture and
  Fisheries
Soil and Water Department
PO Box 1509
Dubai

Jordan

National Resources Authority PO 950260 Amman	Topographic maps, geological maps and information
Jordan Water Authority Amman	Hydrometric records

References and further reading

The reader is also referred to the other Handbooks in this series, which are listed on the inside front cover.

BANNISTER, A. & RAYMOND, S. (1984) *Surveying* (5th edn), Pitman, London/ Marshfield, Massachusetts, 510 pp.

BARNES, J. W. (1981) *Basic Geological Mapping*, Geological Society Handbook, Open University Press, Milton Keynes/Halsted Press, New York/ Toronto, 112 pp.

BRANDON, T. W. (ed.) (1986) *Groundwater: Occurrence, Development and Protection*, Water Practice Manual No. 5, Institution of Water Engineers and Scientists, London, 615 pp.

BRASSINGTON, R. (1983) *Finding Water*, London, Pelham Books, 184 pp.

BRITISH STANDARDS INSTITUTION (1965) *Thin Plate Weirs and Venturi Flumes*, British Standard Code of Practice 3680: Part 4A. London, British Standards Institution, 92 pp.

BRITISH STANDARDS INSTITUTION (1983) *British Standard Code of Practice for Test Pumping Water Wells*, BS 6316: 1983, London, British Standards Institution, 44 pp.

BROWN, R. H., KONOPLYANTSEV, A. A., INESON, J. & KOVALENSKY, V. S. (1983) *Ground-water Studies – An International Guide for Research and Practice*, (incorporating Supplements 1, 2, 3 and 4), Paris, Unesco, 534 pp.

CLARK, L. (1988) *Water Wells and Boreholes*, Geological Society Handbook, Open University Press, Milton Keynes/Halsted Press, New York/ Toronto.

DALY, D. & WRIGHT, G. R. (1982) *Waste Disposal Sites – Geotechnical Guidelines for their Selection, Design and Management*, Geological Survey of Ireland Information Circular 82/1, Dublin, Ministry of Industry and Energy, 50 pp.

DAVIS, S. N. & DEWEIST, R. J. M. (1966) *Hydrogeology*, New York/London/ Sydney, Wiley, 463 pp.

DREW, D. P. & SMITH, D. I. (1969) *Techniques for the Tracing of Subterranean Drainage*, Technical Bulletin No. 2, British Geomorphological Research Group, Norwich, Geo Abstracts, 36 pp.

Field Hydrogeology

DRISCOLL, F. G. (1986) *Groundwater and Wells* (2nd edn), St Paul, Minnesota, Johnson Division, 1089 pp.

FREEZE, R. A. & CHERRY, J. A. (1979) *Groundwater*, Englewood Cliffs, NJ, Prentice-Hall, 604 pp.

GRINDLEY, J. (1969) *The Calculation of Actual Evaporation and Soil Moisture Deficit over Specified Catchment Areas*, Hydrological Memorandum No. 38, London, Meteorological Office, HMSO, 10 pp.

GRINDLEY, J. (1970) 'Estimation and mapping of evaporation', *World Water Balance*, Vol. 1, IASH Publication No. 92, pp. 200–213.

HEAD K. H. (1982) *Manual of Soil Laboratory Testing* (Vols 1 and 2), London/Plymouth, Pentech Press, 748 pp.

HILL, R. A. (1940) Geochemical patterns in Coachell Valley. *Trans. Am. Geophys. Union*, 21, 46–53.

INSTITUTION OF GEOLOGISTS (1985) *The Geologist's Directory* (3rd edn), London, Institution of Geologists, 190 pp.

INSTITUTION OF GEOLOGISTS (1985) *Guidance Notes on Report Writing*, London, Institution of Geologists, 10 pp.

KRUSEMAN, G. P. & DE RIDDER, N. A. (1983) *Analysis and Evaluation of Pumping Test Data* (3rd edn), Bulletin 11, International Institute for Land Reclamation and Development, Wageningen, Netherlands, 200 pp.

MAZOR, E. (forthcoming) *Groundwater Quality and Chemistry*, Geological Society Handbook, Open University Press, Milton Keynes/Halsted Press, New York/Toronto.

METEOROLOGICAL OFFICE (1981) *Handbook of Meteorological Instruments* (2nd edn). Meteorological Office, HMSO, London, 34 pp.

MILSOM, J. S. (1988) *Field Guide to Geophysics*, Geological Society Handbook, Open University Press, Milton Keynes/Halsted Press, New York/Toronto.

NAYLOR, J. A., ROWLAND, C. D., YOUNG, C. P. & BARBER, C. (1978) *The Investigation of Landfill Sites*, Technical Report, TR91, Water Research Centre, Marlow, Bucks, 68 pp.

PENMAN, H. L. (1948) 'Natural evaporation from open water, bare soil and grass', *Proc. R. Soc. Lond.*, 193, 120–145.

PENMAN, H. L. (1950a) 'Evaporation over the British Isles', *Q. J. R. Meteorol. Soc.*, 96, 372–383.

PENMAN, H. L. (1950b) 'The water balance of the Stour catchment area', *J. Inst. Water Eng.*, 4, 457–469.

PIPER, A. M. (1944) 'A graphic procedure in the geochemical interpretation of water analyses', *Trans. Am. Geophys. Union*, 25, 914–928.

PRICE M. (1985) *Introducing Groundwater*, London/Boston/Sydney, George Allen & Unwin, 195 pp.

168

RUSHTON, K. R. & REDSHAW, S. C. (1979) *Seepage and Groundwater Flow*, New York/London/Sydney, John Wiley, 339 pp.

SHAW, E. M. (1983) *Hydrology in Practice*, Wokingham, Van Nostrand Reinhold (UK), 569 pp.

THORNTHWAITE, C. W. (1948) An approach towards a rational classification of climate. *Geog. Rev.*, 38, 55–94.

TODD, D. K. (1980) *Ground Water Hydrology* (2nd edn), New York/London/Sydney, John Wiley, 535 pp.

WAY, D. S. (1973) *Terrain Analysis: a Guide to Site Selection using Aerial Photographic Interpretation*, Strandsburg, PA, Dowden, Hutchison & Ross, 392 pp.

WILSON, E. M. (1974) *Engineering Hydrology* (2nd edn), London, Macmillan, 232 pp.

WRIGHT, G. R. (1985) *Pumping Tests, a Guide to the Testing of Water Wells for Public, Industrial and Farm Supplies*. Geological Survey of Ireland Information Circular 85/2, Ministry of Energy, Dublin, 26 pp.

First aid and survival

ANON. (1980) *First Aid*, St John's Ambulance Association.

LANGMUIR, E. (1987) *Mountain Leadership* (2nd edn), Scottish Sports Council.

RENOUF, J. & HULSE, S. (1978) *First Aid for Hill Walkers and Climbers*, Penguin Books.

Scientific papers

The following international journals contain papers on a wide variety of current hydrogeological topics and should be read by practising hydrogeologists.

Ground Water. Published six times per year by the National Water Well Association, Dublin, Ohio.

The Journal of Hydrology. Published with four issues per volume and four volumes per year by the Elsevier Scientific Publishing Company, Amsterdam, The Netherlands.

The Quarterly Journal of Engineering Geology. Published four times per year by the Geological Society of London.

Index

Index

Index

174